How to Destroy Surveillance Capitalism

Dear Puneet,
Merry X-Mas 2023
Your friend from
Berlin,
Joachim

How to Destroy
Surveillance Capitalism

Cory Doctorow

Published by Stonesong Digital, LLC
New York, NY USA

This is a work of fiction based on reported sightings and incidents. Names, characters, places, and incidents either are the product of the author's imagination or are used fictitiously. Any resemblance to actual persons. living. dead. or undead, events. or locations is entirely coincidental.

HOW TO DESTROY SURVEILLANCE CAPITALISM by Cory Doctorow © 2020 Cory Doctorow

Cover design and illustrations by Shira Inbar

Production management by Stonesong Digital

ebook ISBN: 978-1-7362059-1-4
paperback ISBN: 978-1-7362059-0-7

HOW TO DESTROY SURVEILLANCE CAPITALISM originally appeared on Medium's *OneZero*.

All rights reserved.

10 9 8 7 6 S 4 3 2 1

First Edition, 2020

Contents

The net of a thousand lies   /   1

Digital rights activism, a quarter-century on   /   6

Tech exceptionalism, then and now   /   8

Don't believe the hype   /   10

What is persuasion?   /   12

If data is the new oil, then surveillance capitalism's engine has a leak   /   23

What is Facebook?   /   30

Monopoly and the right to the future tense   /   36

Search order and the right to the future tense   /   42

Monopolists can afford sleeping pills for watchdogs   /   47

Privacy and monopoly   /   56

Ronald Reagan, pioneer of tech monopolism   /   60

Steering with the windshield wipers   /   66

Surveillance still matters   /   68

Dignity and sanctuary   /   72

Afflicting the afflicted   /   75

Any data you collect and retain will eventually leak   /   78

Critical tech exceptionalism is still tech exceptionalism   /   81

How monopolies, not mind control, drive surveillance capitalism: The Snapchat story   /   87

A monopoly over your friends   /   90

Fake news is an epistemological crisis / 96

Tech is different / 104

Ownership of facts / 107

Persuasion works . . . slowly / 111

Paying won't help / 114

An "ecology" moment for trustbusting / 120

Make Big Tech small again / 125

20 GOTO 10 / 128

Up and through / 131

Surveillance Capitalism 'n Kids / 133

The net of a thousand lies

The most surprising thing about the rebirth of flat Earthers in the 21st century is just how widespread the evidence against them is. You can understand how, centuries ago, people who'd never gained a high-enough vantage point from which to see the Earth's curvature might come to the commonsense belief that the flat-seeming Earth was, indeed, flat.

But today, when elementary schools routinely dangle GoPro cameras from balloons and loft them high enough to photograph the Earth's curve—to say nothing of the unexceptional sight of the curved Earth from an airplane window—it takes a heroic effort to maintain the belief that the world is flat.

Likewise for white nationalism and eugenics: In an age where you can become a computational genomics data point by swabbing your cheek and mailing it to a gene-sequencing company along with a modest sum of money, "race science" has never been easier to refute.

We are living through a golden age of both readily available facts and denial of those facts. Terrible ideas that have lingered on the fringes for decades or even centuries have gone mainstream seemingly overnight.

When an obscure idea gains currency, there are only two things that can explain its ascendance: Either the person expressing that idea has gotten a lot better at stating their case, or the proposition has become harder to deny in the face of mounting evidence. In other words, if we want people to take climate change seriously, we can get a bunch of Greta Thunbergs to make eloquent, passionate arguments from podiums, winning our hearts and minds; or we can wait for flood, fire, broiling sun, and pandemics to make the case for us. In practice, we'll probably have to do some of both: The more we're boiling and burning and drowning and wasting away, the easier it will be for the Greta Thunbergs of the world to convince us.

The arguments for ridiculous beliefs in odious conspiracies like anti-vaccination, climate denial, a flat Earth, and eugenics are no better than they were a generation ago. Indeed, they're worse because they are being pitched to people who have at least a background awareness of the refuting facts.

Anti-vax has been around since the first vaccines, but the early anti-vaxxers were pitching people who were less equipped to understand even the most basic ideas from microbiology, and moreover, those people had not witnessed the extermination of mass-murdering diseases like polio, smallpox, and measles. Today's anti-vaxxers are no more eloquent than their forebears, and they have a much harder job.

So can these far-fetched conspiracy theorists really be succeeding on the basis of superior arguments?

Some people think so. Today, there is a widespread belief that machine learning and commercial surveillance can turn even the most fumble-tongued conspiracy theorist into a Svengali who can warp your perceptions and win your belief by locating vulnerable people and then pitching them with A.I.-refined arguments that bypass their rational faculties and turn everyday people into flat Earthers, anti-vaxxers, or even Nazis. When the RAND Corporation blames Facebook for "radicalization" and when Facebook's role in spreading coronavirus misinformation is blamed on its algorithm, the implicit message is that machine learning and surveillance are causing the changes in our consensus about what's true.

After all, in a world where sprawling and incoherent conspiracy theories like Pizzagate and its successor, QAnon, have widespread followings, *something* must be afoot.

But what if there's another explanation? What if it's the material circumstances, and not the arguments, that are making the difference for these conspiracy pitchmen? What if the trauma of living through *real conspiracies* all around us—conspiracies among wealthy people, their lobbyists, and lawmakers to bury inconvenient facts and evidence of wrongdoing (these conspiracies are commonly known as "corruption")—is making people vulnerable to conspiracy theories?

If it's trauma and not contagion—material conditions and not ideology—that is making the difference today and enabling a rise of repulsive misinformation in the face of easily observed facts, that doesn't mean our computer networks are blameless. They're still doing the heavy work of locating vulnerable people and guiding them through a series of ever-more-extreme ideas and communities.

Belief in conspiracy is a raging fire that has done real damage and poses real danger to our planet and species, from epidemics kicked off by vaccine denial to genocides kicked off by racist conspiracies to planetary meltdown caused by denial-inspired climate inaction. Our world is on fire, and so we have to put the fires out—to figure out how to help people see the truth of the world through the conspiracies they've been confused by.

But firefighting is reactive. We need fire *prevention*. We need to strike at the traumatic material conditions that make people vulnerable to the contagion of conspiracy. Here, too, tech has a role to play.

There's no shortage of proposals to address this. From the EU's Terrorist Content Regulation, which requires platforms to police and remove "extremist" content, to the U.S. proposals to force tech companies to spy on their users and hold them liable for their users' bad speech, there's a lot of energy to force tech companies to solve the problems they created.

There's a critical piece missing from the debate, though. All these solutions assume that tech companies are a fixture, that their dominance over the internet is a permanent fact. Proposals to replace Big Tech with a more diffused, pluralistic internet are nowhere to be found. Worse: The "solutions" on the table today *require* Big Tech to stay big because only the very largest companies can afford to implement the systems these laws demand.

Figuring out what we want our tech to look like is crucial if we're going to get out of this mess. Today, we're at a crossroads where we're trying to figure out if we want to fix the Big Tech companies that dominate our internet or if we want to fix the internet itself by unshackling it from Big Tech's stranglehold. We can't do both, so we have to choose.

I want us to choose wisely. Taming Big Tech is integral to fixing the internet, and for that, we need digital rights activism.

Digital rights activism, a quarter-century on

Digital rights activism is more than 30 years old now. The Electronic Frontier Foundation turned 30 this year; the Free Software Foundation launched in 1985. For most of the history of the movement, the most prominent criticism leveled against it was that it was irrelevant: The real activist causes were real-world causes (think of the skepticism when Finland declared broadband a human right in 2010), and real-world activism was shoe-leather activism (think of Malcolm Gladwell's contempt for "clicktivism"). But as tech has grown more central to our daily lives, these accusations of irrelevance have given way first to accusations of insincerity ("You only care about tech because you're shilling for tech companies") to accusations of negligence ("Why didn't you foresee that tech could be such a destructive force?"). But digital rights activism is right where it's always been: looking out for the humans in a world where tech is inexorably taking over.

The latest version of this critique comes in the form of "surveillance capitalism," a term coined by business professor Shoshana Zuboff in her long and influential 2019 book *The Age of Surveillance Capitalism: The Fight for a Human Future at the New Frontier of Power*. Zuboff argues that "surveillance capitalism" is a unique creature of the tech industry and that it is unlike any other abusive commercial practice in history, one that is "constituted by unexpected and often illegible mechanisms of extraction, commodification, and control that effectively exile persons from their own behavior while producing new markets of behavioral prediction and modification. Surveillance capitalism challenges democratic norms and departs in key ways from the centuries-long evolution of market capitalism." It is a new and deadly form of capitalism, a "rogue capitalism," and our lack of understanding of its unique capabilities and dangers represents an existential, species-wide threat. She's right that capitalism today threatens our species, and she's right that tech poses unique challenges to our species and civilization, but she's really wrong about how tech is different and why it threatens our species.

What's more, I think that her incorrect diagnosis will lead us down a path that ends up making Big Tech stronger, not weaker. We need to take down Big Tech, and to do that, we need to start by correctly identifying the problem.

Tech exceptionalism, then and now

Early critics of the digital rights movement—perhaps best represented by campaigning organizations like the Electronic Frontier Foundation, the Free Software Foundation, Public Knowledge, and others that focused on preserving and enhancing basic human rights in the digital realm—damned activists for practicing "tech exceptionalism." Around the turn of the millennium, serious people ridiculed any claim that tech policy mattered in the "real world." Claims that tech rules had implications for speech, association, privacy, search and seizure, and fundamental rights and equities were treated as ridiculous, an elevation of the concerns of sad nerds arguing about *Star Trek* on bulletin board systems above the struggles of the Freedom Riders, Nelson Mandela, or the Warsaw Ghetto Uprising.

In the decades since, accusations of "tech exceptionalism" have only sharpened as tech's role in everyday life has expanded: Now that tech has infiltrated every corner of our life and our online lives have been monopolized by a handful of giants, defenders of digital freedoms are accused of carrying water for Big Tech, providing cover for its self-interested negligence (or worse, nefarious plots).

From my perspective, the digital rights movement has remained stationary while the rest of the world has moved. From the earliest days, the movement's concern was users and the toolsmiths who provided the code they needed to realize their fundamental rights. Digital rights activists only cared about companies to the extent that companies were acting to uphold users' rights (or, just as often, when companies were acting so foolishly that they threatened to bring down new rules that would also make it harder for good actors to help users).

The "surveillance capitalism" critique recasts the digital rights movement in a new light again: not as alarmists who overestimate the importance of their shiny toys nor as shills for Big Tech but as serene deck-chair rearrangers whose long-standing activism is a liability because it makes them incapable of perceiving novel threats as they continue to fight the last century's tech battles.

But tech exceptionalism is a sin no matter who practices it.

Don't believe the hype

You've probably heard that "if you're not paying for the product, you're the product." As we'll see below, that's true, if incomplete. But what is *absolutely* true is that ad-driven Big Tech's customers are advertisers, and what companies like Google and Facebook sell is their ability to convince *you* to buy stuff. Big Tech's product is persuasion. The services—social media, search engines, maps, messaging, and more—are delivery systems for persuasion.

The fear of surveillance capitalism starts from the (correct) presumption that everything Big Tech says about itself is probably a lie. But the surveillance capitalism critique makes an exception for the claims Big Tech makes in its sales literature—the breathless hype in the pitches to potential advertisers online and in ad-tech seminars about the efficacy of its products: It assumes that Big Tech is as good at influencing us as they claim they are when they're selling influencing products to credulous customers. That's a mistake because sales literature is not a reliable indicator of a product's efficacy.

Surveillance capitalism assumes that because advertisers buy a lot of what Big Tech is selling, Big Tech must be selling something real. But Big Tech's massive sales could just as easily be the result of a popular delusion or something even more pernicious: monopolistic control over our communications and commerce.

Being watched changes your behavior, and not for the better. It creates risks for our social progress. Zuboff's book features beautifully wrought explanations of these phenomena. But Zuboff also claims that surveillance literally robs us of our free will—that when our personal data is mixed with machine learning, it creates a system of persuasion so devastating that we are helpless before it. That is, Facebook uses an algorithm to analyze the data it nonconsensually extracts from your daily life and uses it to customize your feed in ways that get you to buy stuff. It is a mind-control ray out of a 1950s comic book, wielded by mad scientists whose supercomputers guarantee them perpetual and total world domination.

What is persuasion?

To understand why you shouldn't worry about mind-control rays—but why you *should* worry about surveillance *and* Big Tech—we must start by unpacking what we mean by "persuasion."

Google, Facebook, and other surveillance capitalists promise their customers (the advertisers) that if they use machine-learning tools trained on unimaginably large data sets of nonconsensually harvested personal information, they will be able to uncover ways to bypass the rational faculties of the public and direct their behavior, creating a stream of purchases, votes, and other desired outcomes.

But there's little evidence that this is happening. Instead, the predictions that surveillance capitalism delivers to its customers are much less impressive. Rather than finding ways to bypass our rational faculties, surveillance capitalists like Mark Zuckerberg mostly do one or more of three things:

1. Segmenting

If you're selling diapers, you have better luck if you pitch them to people in maternity wards. Not everyone who enters or leaves a maternity ward just had a baby, and not everyone who just had a baby is in the market for diapers. But having a baby is a really reliable correlate of being in the market for diapers, and being in a maternity ward is highly correlated with having a baby. Hence diaper ads around maternity wards (and even pitchmen for baby products, who haunt maternity wards with baskets full of freebies).

Surveillance capitalism is segmenting times a billion. Diaper vendors can go way beyond people in maternity wards (though they can do that, too, with things like location-based mobile ads). They can target you based on whether you're reading articles about child-rearing, diapers, or a host of other subjects, and data mining can suggest unobvious keywords to advertise against. They can target you based on the articles you've recently read. They can target you based on what you've recently purchased. They can target you based on whether you receive emails or private messages about these subjects—or even if you speak aloud about them (though Facebook and the like convincingly claim that's not happening—yet).

This is seriously creepy.

But it's not mind control.

It doesn't deprive you of your free will. It doesn't trick you.

Think of how surveillance capitalism works in politics. Surveillance capitalist companies sell political operatives the power to locate people who might be receptive to their pitch. Candidates campaigning on finance industry corruption seek people struggling with debt; candidates campaigning on xenophobia seek out racists. Political operatives have always targeted their message whether their intentions were honorable or not: Union organizers set up pitches at factory gates, and white supremacists hand out fliers at John Birch Society meetings.

But this is an inexact and thus wasteful practice. The union organizer can't know which worker to approach on the way out of the factory gates and may waste their time on a covert John Birch Society member; the white supremacist doesn't know which of the Birchers are so delusional that making it to a meeting is as much as they can manage and which ones might be convinced to cross the country to carry a tiki torch through the streets of Charlottesville, Virginia.

Because targeting improves the yields on political pitches, it can accelerate the pace of political upheaval by making it possible for everyone who has secretly wished for the toppling of an autocrat—or just an 11-term incumbent politician—to find everyone else who feels the same way at very low cost. This has been critical to the rapid crystallization of recent political movements including Black Lives Matter and Occupy Wall Street as well as less savory players like the far-right white nationalist movements that marched in Charlottesville.

It's important to differentiate this kind of political organizing from influence campaigns; finding people who secretly agree with you isn't the same as convincing people to agree with you. The rise of phenomena like nonbinary or otherwise nonconforming gender identities is often characterized by reactionaries as the result of online brainwashing campaigns that convince impressionable people that they have been secretly queer all along.

But the personal accounts of those who have come out tell a different story where people who long harbored a secret about their gender were emboldened by others coming forward and where people who knew that they were different but lacked a vocabulary for discussing that difference learned the right words from these low-cost means of finding people and learning about their ideas.

2. Deception

Lies and fraud are pernicious, and surveillance capitalism supercharges them through targeting. If you want to sell a fraudulent payday loan or subprime mortgage, surveillance capitalism can help you find people who are both desperate and unsophisticated and thus receptive to your pitch. This accounts for the rise of many phenomena, like multilevel marketing schemes, in which deceptive claims about potential earnings and the efficacy of sales techniques are targeted at desperate people by advertising against search queries that indicate, for example, someone struggling with ill-advised loans.

Surveillance capitalism also abets fraud by making it easy to locate other people who have been similarly deceived, forming a community of people who reinforce one another's false beliefs. Think of the forums in which people who are being victimized by multilevel marketing frauds gather to trade tips on how to improve their luck in peddling the product.

Sometimes, online deception involves replacing someone's correct beliefs with incorrect ones, as it does in the anti-vaccination movement, whose victims are often people who start out believing in vaccines but are convinced by seemingly plausible evidence that leads them into the false belief that vaccines are harmful.

But it's much more common for fraud to succeed when it doesn't have to displace a true belief. When my daughter contracted head lice at daycare, one of the daycare workers told me I could get rid of them by treating her hair and scalp with olive oil. I didn't know anything about head lice, and I assumed that the daycare worker did, so I tried it (it didn't work, and it doesn't work). It's easy to end up with false beliefs when you simply don't know any better and when those beliefs are conveyed by someone who seems to know what they're doing.

This is pernicious and difficult—and it's also the kind of thing the internet can help guard against by making true information available, especially in a form that exposes the underlying deliberations among parties with sharply divergent views, such as Wikipedia. But it's not brainwashing; it's fraud. In the majority of cases, the victims of these fraud campaigns have an informational void filled in the customary way, by consulting a seemingly reliable source. If I look up the length of the Brooklyn Bridge and learn that it is 5,800 feet long, but in reality, it is 5,989 feet long; the underlying deception is a problem, but it's a problem with a simple remedy. It's a very different problem from the anti-vax issue in which someone's true belief is displaced by a false one by means of sophisticated persuasion.

3. Domination

Surveillance capitalism is the result of monopoly. Monopoly is the cause, and surveillance capitalism and its negative outcomes are the effects of monopoly. I'll get into this in depth later, but for now, suffice it to say that the tech industry has grown up with a radical theory of antitrust that has allowed companies to grow by merging with their rivals, buying up their nascent competitors, and expanding to control whole market verticals.

One example of how monopolism aids in persuasion is through dominance: Google makes editorial decisions about its algorithms that determine the sort order of the responses to our queries. If a cabal of fraudsters have set out to trick the world into thinking that the Brooklyn Bridge is 5,800 feet long, and if Google gives a high search rank to this group in response to queries like "How long is the Brooklyn Bridge?" then the first 8 or 10 screens' worth of Google results could be wrong. And since most people don't go beyond the first couple of results—let alone the first *page* of results—Google's choice means that many people will be deceived.

Google's dominance over search—more than 86% of web searches are performed through Google—means that the way it orders its search results has an outsized effect on public beliefs. Ironically, Google claims this is why it can't afford to have any transparency in its algorithm design: Google's search dominance makes the results of its sorting too important to risk telling the world how it arrives at those results lest some bad actor discover a flaw in the ranking system and exploit it to push its point of view to the top of the search results. There's an obvious remedy to a company that is too big to audit: Break it up into smaller pieces.

Zuboff calls surveillance capitalism a "rogue capitalism" whose data-hoarding and machine-learning techniques rob us of our free will. But influence campaigns that seek to displace existing, correct beliefs with false ones have an effect that is small and temporary while monopolistic dominance over informational systems has massive, enduring effects. Controlling the results to the world's search queries means controlling access both to arguments and their rebuttals and, thus, control over much of the world's beliefs. If our concern is how corporations are foreclosing on our ability to make up our own minds and determine our own futures, the impact of dominance far exceeds the impact of manipulation and should be central to our analysis and any remedies we seek.

4. Bypassing our rational faculties

This is the good stuff: using machine learning, "dark patterns," engagement hacking, and other techniques to get us to do things that run counter to our better judgment. This is mind control.

Some of these techniques have proven devastatingly effective (if only in the short term). The use of countdown timers on a purchase completion page can create a sense of urgency that causes you to ignore the nagging internal voice suggesting that you should shop around or sleep on your decision. The use of people from your social graph in ads can provide "social proof" that a purchase is worth making. Even the auction system pioneered by eBay is calculated to play on our cognitive blind spots, letting us feel like we "own" something because we bid on it, thus encouraging us to bid again when we are outbid to ensure that "our" things stay ours.

Games are extraordinarily good at this. "Free-to-play" games manipulate us through many techniques, such as presenting players with a series of smoothly escalating challenges that create a sense of mastery and accomplishment but which sharply transition into a set of challenges that are impossible to overcome without paid upgrades. Add some social proof to the mix—a stream of notifications about how well your friends are faring—and before you know it, you're buying virtual power-ups to get to the next level.

Companies have risen and fallen on these techniques, and the "fallen" part is worth paying attention to. In general, living things adapt to stimulus: Something that is very compelling or noteworthy when you first encounter it fades with repetition until you stop noticing it altogether. Consider the refrigerator hum that irritates you when it starts up but disappears into the background so thoroughly that you only notice it when it stops again.

That's why behavioral conditioning uses "intermittent reinforcement schedules." Instead of giving you a steady drip of encouragement or setbacks, games and gamified services scatter rewards on a randomized schedule—often enough to keep you interested and random enough that you can never quite find the pattern that would make it boring.

Intermittent reinforcement is a powerful behavioral tool, but it also represents a collective action problem for surveillance capitalism. The "engagement techniques" invented by the behaviorists of surveillance capitalist companies are quickly copied across the whole sector so that what starts as a mysteriously compelling fillip in the design of a service—like "pull to refresh" or alerts when someone likes your posts or side quests that your characters get invited to while in the midst of main quests—quickly becomes dully ubiquitous. The impossible-to-nail-down nonpattern of randomized drips from your phone becomes a noise wall of sound as every single app and site starts to make use of whatever seems to be working at the time.

From the surveillance capitalist's point of view, our adaptive capacity is like a harmful bacterium that deprives it of its food source—our attention—and novel techniques for snagging that attention are like new antibiotics that can be used to breach our defenses and destroy our self-determination. And there *are* techniques like that. Who can forget the Great Zynga Epidemic, when all of our friends were caught in *FarmVille*'s endless, mindless dopamine loops? But every new attention-commanding technique is jumped on by the whole industry and used so indiscriminately that antibiotic resistance sets in. Given enough repetition, almost all of us develop immunity to even the most powerful techniques—by 2013, two years after Zynga's peak, its user base had halved.

Not everyone, of course. Some people never adapt to stimulus, just as some people never stop hearing the hum of the refrigerator. This is why most people who are exposed to slot machines play them for a while and then move on while a small and tragic minority liquidate their kids' college funds, buy adult diapers, and position themselves in front of a machine until they collapse.

But surveillance capitalism's margins on behavioral modification suck. Tripling the rate at which someone buys a widget sounds great unless the base rate is way less than 1% with an improved rate of . . . still less than 1%. Even penny slot machines pull down pennies for every spin while surveillance capitalism rakes in infinitesimal penny fractions.

Slot machines' high returns mean that they can be profitable just by draining the fortunes of the small rump of people who are pathologically vulnerable to them and unable to adapt to their tricks. But surveillance capitalism can't survive on the fractional pennies it brings down from that vulnerable sliver—that's why, after the Great Zynga Epidemic had finally burned itself out, the small number of still-addicted players left behind couldn't sustain it as a global phenomenon. And new powerful attention weapons aren't easy to find, as is evidenced by the long years since the last time Zynga had a hit. Despite the hundreds of millions of dollars that Zynga has to spend on developing new tools to blast through our adaptation, it has never managed to repeat the lucky accident that let it snag so much of our attention for a brief moment in 2009. Powerhouses like Supercell have fared a little better, but they are rare and throw away many failures for every success.

The vulnerability of small segments of the population to dramatic, efficient corporate manipulation is a real concern that's worthy of our attention and energy. But it's not an existential threat to society.

If data is the new oil, then surveillance capitalism's engine has a leak

This adaptation problem offers an explanation for one of surveillance capitalism's most alarming traits: its relentless hunger for data and its endless expansion of data-gathering capabilities through the spread of sensors, online surveillance, and acquisition of data streams from third parties.

Zuboff observes this phenomenon and concludes that data must be very valuable if surveillance capitalism is so hungry for it. (In her words: "Just as industrial capitalism was driven to the continuous intensification of the means of production, so surveillance capitalists and their market players are now locked into the continuous intensification of the means of behavioral modification and the gathering might of instrumentarian power.") But what if the voracious appetite is because data has such a short half-life—because people become inured so quickly to new, data-driven persuasion techniques—that the companies are locked in an arms race with our limbic system? What if it's all a Red Queen's race where they have to run ever faster—collect ever-more data—just to stay in the same spot?

Of course, all of Big Tech's persuasion techniques work in concert with one another, and collecting data is useful beyond mere behavioral trickery.

If someone wants to recruit you to buy a refrigerator or join a pogrom, they might use profiling and targeting to send messages to people they judge to be good sales prospects. The messages themselves may be deceptive, making claims about things you're not very knowledgeable about (food safety and energy efficiency or eugenics and historical claims about racial superiority). They might use search engine optimization and/or armies of fake reviewers and commenters and/or paid placement to dominate the discourse so that any search for further information takes you back to their messages. And finally, they may refine the different pitches using machine learning and other techniques to figure out what kind of pitch works best on someone like you.

Each phase of this process benefits from surveillance: The more data they have, the more precisely they can profile you and target you with specific messages. Think of how you'd sell a fridge if you knew that the warranty on your prospect's fridge just expired and that they were expecting a tax rebate in April.

Also, the more data they have, the better they can craft deceptive messages—if I know that you're into genealogy,

I might not try to feed you pseudoscience about genetic differences between "races," sticking instead to conspiratorial secret histories of "demographic replacement" and the like.

Facebook also helps you locate people who have the same odious or antisocial views as you. It makes it possible to find other people who want to carry tiki torches through the streets of Charlottesville in Confederate cosplay. It can help you find other people who want to join your militia and go to the border to look for undocumented migrants to terrorize. It can help you find people who share your belief that vaccines are poison and that the Earth is flat.

There is one way in which targeted advertising uniquely benefits those advocating for socially unacceptable causes: It is invisible. Racism is widely geographically dispersed, and there are few places where racists—and only racists—gather. This is similar to the problem of selling refrigerators in that potential refrigerator purchasers are geographically dispersed and there are few places where you can buy an ad that will be primarily seen by refrigerator customers. But buying a refrigerator is socially acceptable while being a Nazi is not, so you can buy a billboard or advertise in the newspaper sports section for your refrigerator business, and the only potential downside is that your ad will be seen by a lot of people who don't want refrigerators, resulting in a lot of wasted expense.

But even if you wanted to advertise your Nazi movement on a billboard or prime-time TV or the sports section, you would struggle to find anyone willing to sell you the space for your ad partly because they disagree with your views and partly because they fear censure (boycott, reputational damage, etc.) from other people who disagree with your views.

Targeted ads solve this problem: On the internet, every ad unit can be different for every person, meaning that you can buy ads that are only shown to people who appear to be Nazis and not to people who hate Nazis. When there's spillover—when someone who hates racism is shown a racist recruiting ad— there is some fallout; the platform or publication might get an angry public or private denunciation. But the nature of the risk assumed by an online ad buyer is different than the risks to a traditional publisher or billboard owner who might want to run a Nazi ad.

Online ads are placed by algorithms that broker between a diverse ecosystem of self-serve ad platforms that anyone can buy an ad through, so the Nazi ad that slips into your favorite online publication isn't seen as their moral failing but rather as a failure in some distant, upstream ad supplier. When a publication gets a complaint about an offensive ad that's appearing in one of its units, it can take some steps to block that ad, but the Nazi might buy a slightly different ad from a different broker serving the same unit. And in any event, internet users increasingly understand that when they see an ad, it's likely that the advertiser did not choose that publication and that the publication has no idea who its advertisers are.

These layers of indirection between advertisers and publishers serve as moral buffers: Today's moral consensus is largely that publishers shouldn't be held responsible for the ads that appear on their pages because they're not actively choosing to put those ads there. Because of this, Nazis are able to overcome significant barriers to organizing their movement.

Data has a complex relationship with domination. Being able to spy on your customers can alert you to their preferences for your rivals and allow you to head off your rivals at the pass.

More importantly, if you can dominate the information space while also gathering data, then you make other deceptive tactics stronger because it's harder to break out of the web of deceit you're spinning. Domination—that is, ultimately becoming a monopoly—and not the data itself is the supercharger that makes every tactic worth pursuing because monopolistic domination deprives your target of an escape route.

If you're a Nazi who wants to ensure that your prospects primarily see deceptive, confirming information when they search for more, you can improve your odds by seeding the search terms they use through your initial communications. You don't need to own the top 10 results for "voter suppression" if you can convince your marks to confine their search terms to "voter fraud," which throws up a very different set of search results.

Surveillance capitalists are like stage mentalists who claim that their extraordinary insights into human behavior let them guess the word that you wrote down and folded up in your pocket but who really use shills, hidden cameras, sleight of hand, and brute-force memorization to amaze you.

Or perhaps they're more like pick-up artists, the misogynistic cult that promises to help awkward men have sex with women by teaching them "neurolinguistic programming" phrases, body language techniques, and psychological manipulation tactics like "negging"—offering unsolicited negative feedback to women to lower their self-esteem and prick their interest.

Some pick-up artists eventually manage to convince women to go home with them, but it's not because these men have figured out how to bypass women's critical faculties. Rather, pick-up artists' "success" stories are a mix of women who were incapable of giving consent, women who were coerced, women who were intoxicated, self-destructive women, and a few women who were sober and in command of their faculties but who didn't realize straightaway that they were with terrible men but rectified the error as soon as they could.

Pick-up artists *believe* they have figured out a secret back door that bypasses women's critical faculties, but they haven't. Many of the tactics they deploy, like negging, became the butt of jokes (just like people joke about bad ad targeting), and there's a good chance that anyone they try these tactics on will immediately recognize them and dismiss the men who use them as irredeemable losers.

Pick-up artists are proof that people can believe they have developed a system of mind control *even when it doesn't work*. Pick-up artists simply exploit the fact that one-in-a-million chances can come through for you if you make a million attempts, and then they assume that the other 999,999 times, they simply performed the technique incorrectly and commit themselves to doing better next time. There's only one group of people who find pick-up-artist lore reliably convincing: other would-be pick-up artists whose anxiety and insecurity make them vulnerable to scammers and delusional men who convince them that if they pay for tutelage and follow instructions, then they will someday succeed. Pick-up artists assume they fail to entice women because they are bad at being pick-up artists, not because pick-up artistry is bullshit. Pick-up artists are bad at selling themselves to women, but they're much better at selling themselves to men who pay to learn the secrets of pick-up artistry.

Department store pioneer John Wanamaker is said to have lamented, "Half the money I spend on advertising is wasted; the trouble is I don't know which half." The fact that Wanamaker thought that only half of his advertising spending was wasted is a tribute to the persuasiveness of advertising executives, who are *much* better at convincing potential clients to buy their services than they are at convincing the general public to buy their clients' wares.

What is Facebook?

Facebook is heralded as the origin of all of our modern plagues, and it's not hard to see why. Some tech companies want to lock their users in but make their money by monopolizing access to the market for apps for their devices and gouging them on prices rather than by spying on them (like Apple). Some companies don't care about locking in users because they've figured out how to spy on them no matter where they are and what they're doing and can turn that surveillance into money (Google). Facebook alone among the Western tech giants has built a business based on locking in its users *and* spying on them all the time.

Facebook's surveillance regime is really without parallel in the Western world. Though Facebook tries to prevent itself from being visible on the public web, hiding most of what goes on there from people unless they're logged into Facebook, the company has nevertheless booby-trapped the entire web with surveillance tools in the form of Facebook "Like" buttons that web publishers include on their sites to boost their Facebook profiles. Facebook also makes various libraries and other useful code snippets available to web publishers that act as surveillance tendrils on the sites where they're used, funneling information about visitors to the site—newspapers, dating sites, message boards—to Facebook.

Facebook offers similar tools to app developers, so the apps—games, fart machines, business-review services, apps for keeping abreast of your kid's schooling—you use will send information about your activities to Facebook even if you don't have a Facebook account and even if you don't download or use Facebook apps. On top of all that, Facebook buys data from third-party brokers on shopping habits, physical location, use of "loyalty" programs, financial transactions, etc., and cross-references that with the dossiers it develops on activity on Facebook and with apps and the public web.

Though it's easy to integrate the web with Facebook—linking to news stories and such—Facebook products are generally not available to be integrated back into the web itself. You can embed a tweet in a Facebook post, but if you embed a Facebook post in a tweet, you just get a link back to Facebook and must log in before you can see it. Facebook has used extreme technological and legal countermeasures to prevent rivals from allowing their users to embed Facebook snippets in competing services or to create alternative interfaces to Facebook that merge your Facebook inbox with those of other services that you use.

And Facebook is incredibly popular, with 2.3 billion claimed users (though many believe this figure to be inflated). Facebook has been used to organize genocidal pogroms, racist riots, anti-vaccination movements, flat Earth cults, and the political lives of some of the world's ugliest, most brutal autocrats. There are some really alarming things going on in the world, and Facebook is implicated in many of them, so it's easy to conclude that these bad things are the result of Facebook's mind-control system, which it rents out to anyone with a few bucks to spend.

To understand what role Facebook plays in the formulation and mobilization of antisocial movements, we need to understand the dual nature of Facebook.

Because it has a lot of users and a lot of data about those users, Facebook is a very efficient tool for locating people with hard-to-find traits, the kinds of traits that are widely diffused in the population such that advertisers have historically struggled to find a cost-effective way to reach them. Think back to refrigerators: Most of us only replace our major appliances a few times in our entire lives. If you're a refrigerator manufacturer or retailer, you have these brief windows in the life of a consumer during which they are pondering a purchase, and you have to somehow reach them. Anyone who's ever registered a title change after buying a house can attest that appliance manufacturers are incredibly desperate to reach anyone who has even the slenderest chance of being in the market for a new fridge.

Facebook makes finding people shopping for refrigerators a *lot* easier. It can target ads to people who've registered a new home purchase, to people who've searched for refrigerator buying advice, to people who have complained about their fridge dying, or any combination thereof. It can even target people who've recently bought *other* kitchen appliances on the theory that someone who's just replaced their stove and dishwasher might be in a fridge-buying kind of mood. The vast majority of people who are reached by these ads will not be in the market for a new fridge, but—crucially—the percentage of people who *are* looking for fridges that these ads reach is *much* larger than it is than for any group that might be subjected to traditional, offline targeted refrigerator marketing.

Facebook also makes it a lot easier to find people who have the same rare disease as you, which might have been impossible in earlier eras—the closest fellow sufferer might otherwise be hundreds of miles away. It makes it easier to find people who went to the same high school as you even though decades have passed and your former classmates have all been scattered to the four corners of the Earth.

Facebook also makes it much easier to find people who hold the same rare political beliefs as you. If you've always harbored a secret affinity for socialism but never dared utter this aloud lest you be demonized by your neighbors, Facebook can help you discover other people who feel the same way (and it might just demonstrate to you that your affinity is more widespread than you ever suspected). It can make it easier to find people who share your sexual identity. And again, it can help you to understand that what you thought was a shameful secret that affected only you was really a widely shared trait, giving you both comfort and the courage to come out to the people in your life.

All of this presents a dilemma for Facebook: Targeting makes the company's ads more effective than traditional ads, but it also lets advertisers see just how effective their ads are. While advertisers are pleased to learn that Facebook ads are more effective than ads on systems with less sophisticated targeting, advertisers can also see that in nearly every case, the people who see their ads ignore them. Or, at best, the ads work on a subconscious level, creating nebulous unmeasurables like "brand recognition." This means that the price per ad is very low in nearly every case.

To make things worse, many Facebook groups spark precious little discussion. Your little-league soccer team, the people with the same rare disease as you, and the people you share a political affinity with may exchange the odd flurry of messages at critical junctures, but on a daily basis, there's not much to say to your old high school chums or other hockey-card collectors.

With nothing but "organic" discussion, Facebook would not generate enough traffic to sell enough ads to make the money it needs to continually expand by buying up its competitors while returning handsome sums to its investors.

So Facebook has to gin up traffic by sidetracking its own forums: Every time Facebook's algorithm injects controversial materials—inflammatory political articles, conspiracy theories, outrage stories—into a group, it can hijack that group's nominal purpose with its desultory discussions and supercharge those discussions by turning them into bitter, unproductive arguments that drag on and on. Facebook is optimized for engagement, not happiness, and it turns out that automated systems are pretty good at figuring out things that people will get angry about.

Facebook *can* modify our behavior but only in a couple of trivial ways. First, it can lock in all your friends and family members so that you check and check and check with Facebook to find out what they are up to; and second, it can make you angry and anxious. It can force you to choose between being interrupted constantly by updates—a process that breaks your concentration and makes it hard to be introspective—and staying in touch with your friends. This is a very limited form of mind control, and it can only really make us miserable, angry, and anxious.

This is why Facebook's targeting systems—both the ones it shows to advertisers and the ones that let users find people who share their interests—are so next-gen and smooth and easy to use as well as why its message boards have a toolset that seems like it hasn't changed since the mid-2000s. If Facebook delivered an equally flexible, sophisticated message-reading system to its users, those users could defend themselves against being nonconsensually eyeball-fucked with Donald Trump headlines.

The more time you spend on Facebook, the more ads it gets to show you. The solution to Facebook's ads only working one in a thousand times is for the company to try to increase how much time you spend on Facebook by a factor of a thousand. Rather than thinking of Facebook as a company that has figured out how to show you exactly the right ad in exactly the right way to get you to do what its advertisers want, think of it as a company that has figured out how to make you slog through an endless torrent of arguments even though they make you miserable, spending so much time on the site that it eventually shows you at least one ad that you respond to.

Monopoly and the right to the future tense

Zuboff and her cohort are particularly alarmed at the extent to which surveillance allows corporations to influence our decisions, taking away something she poetically calls "the right to the future tense"—that is, the right to decide for yourself what you will do in the future.

It's true that advertising can tip the scales one way or another: When you're thinking of buying a fridge, a timely fridge ad might end the search on the spot. But Zuboff puts enormous and undue weight on the persuasive power of surveillance-based influence techniques. Most of these don't work very well, and the ones that do won't work for very long. The makers of these influence tools are confident they will someday refine them into systems of total control, but they are hardly unbiased observers, and the risks from their dreams coming true are very speculative.

By contrast, Zuboff is rather sanguine about 40 years of lax antitrust practice that has allowed a handful of companies to dominate the internet, ushering in an information age with, as one person on Twitter noted, five giant websites each filled with screenshots of the other four.

However, if we are to be alarmed that we might lose the right to choose for ourselves what our future will hold, then monopoly's nonspeculative, concrete, here-and-now harms should be front and center in our debate over tech policy.

Start with "digital rights management." In 1998, Bill Clinton signed the Digital Millennium Copyright Act (DMCA) into law. It's a complex piece of legislation with many controversial clauses but none more so than Section 1201, the "anti-circumvention" rule.

This is a blanket ban on tampering with systems that restrict access to copyrighted works. The ban is so thoroughgoing that it prohibits removing a copyright lock even when no copyright infringement takes place. This is by design: The activities that the DMCA's Section 1201 sets out to ban are not copyright infringements; rather, they are legal activities that frustrate manufacturers' commercial plans.

For example, Section 1201's first major application was on DVD players as a means of enforcing the region coding built into those devices. DVD-CCA, the body that standardized DVDs and DVD players, divided the world into six regions and specified that DVD players must check each disc to determine which regions it was authorized to be played in. DVD players would have their own corresponding region (a DVD player bought in the United States would be region 1 while one bought in India would be region 5). If the player and the disc's region matched, the player would play the disc; otherwise, it would reject it.

However, watching a lawfully produced disc in a country other than the one where you purchased it is not copyright infringement—it's the opposite. Copyright law imposes this duty on customers for a movie: You must go into a store, find a licensed disc, and pay the asking price. Do that—and *nothing else*—and you and copyright are square with one another.

The fact that a movie studio wants to charge Indians less than Americans or release in Australia later than it releases in the United Kingdom has no bearing on copyright law. Once you lawfully acquire a DVD, it is no copyright infringement to watch it no matter where you happen to be.

So DVD and DVD player manufacturers would not be able to use accusations of abetting copyright infringement to punish manufacturers who made noncompliant players that would play discs from any region or repair shops that modified players to let you watch out-of-region discs or software programmers who created programs to let you do this.

That's where Section 1201 of the DMCA comes in: By banning tampering with an "access control," the rule gave manufacturers and rights holders standing to sue competitors who released superior products with lawful features that the market demanded (in this case, region-free players).

This is an odious scam against consumers, but as time went by, Section 1201 grew to encompass a rapidly expanding constellation of devices and services as canny manufacturers have realized certain things:

- Any device with software in it contains a "copyrighted work"—i.e., the software.

- A device can be designed so that reconfiguring the software requires bypassing an "access control for copyrighted works," which is a potential felony under Section 1201.

- Thus, companies can control their customers' behavior after they take home their purchases by designing products so that all unpermitted uses require modifications that fall afoul of Section 1201.

Section 1201 then becomes a means for manufacturers of all descriptions to force their customers to arrange their affairs to benefit the manufacturers' shareholders instead of themselves.

This manifests in many ways: from a new generation of inkjet printers that use countermeasures to prevent third-party ink that cannot be bypassed without legal risks to similar systems in tractors that prevent third-party technicians from swapping in the manufacturer's own parts that are not recognized by the tractor's control system until it is supplied with a manufacturer's unlock code.

Closer to home, Apple's iPhones use these measures to prevent both third-party service and third-party software installation.

This allows Apple to decide when an iPhone is beyond repair and must be shredded and landfilled as opposed to the iPhone's purchaser. (Apple is notorious for its environmentally catastrophic policy of destroying old electronics rather than permitting them to be cannibalized for parts.) This is a very useful power to wield, especially in light of CEO Tim Cook's January 2019 warning to investors that the company's profits are endangered by customers choosing to hold their phones for longer rather than replacing them.

Apple's use of copyright locks also allows it to establish a monopoly over how its customers acquire software for their mobile devices. The App Store's commercial terms guarantee Apple a share of all revenues generated by the apps sold there, meaning that Apple gets paid when you buy an app from its store and then continues to get paid every time you buy something using that app. This comes out of the bottom line of software developers, who must either charge more or accept lower profits for their products.

Crucially, Apple's use of copyright locks gives it the power to make editorial decisions about which apps you may and may not install on your own device. Apple has used this power to reject dictionaries for containing obscene words; to limit political speech, especially from apps that make sensitive political commentary such as an app that notifies you every time a U.S. drone kills someone somewhere in the world; and to object to a game that commented on the Israel-Palestine conflict.

Apple often justifies monopoly power over software installation in the name of security, arguing that its vetting of apps for its store means that it can guard its users against apps that contain surveillance code. But this cuts both ways. In China, the government ordered Apple to prohibit the sale of privacy tools like VPNs with the exception of VPNs that had deliberately introduced flaws designed to let the Chinese state eavesdrop on users. Because Apple uses technological countermeasures—with legal backstops—to block customers from installing unauthorized apps, Chinese iPhone owners cannot readily (or legally) acquire VPNs that would protect them from Chinese state snooping.

Zuboff calls surveillance capitalism a "rogue capitalism." Theoreticians of capitalism claim that its virtue is that it aggregates information in the form of consumers' decisions, producing efficient markets. Surveillance capitalism's supposed power to rob its victims of their free will through computationally supercharged influence campaigns means that our markets no longer aggregate customers' decisions because we customers no longer decide—we are given orders by surveillance capitalism's mind-control rays.

If our concern is that markets cease to function when consumers can no longer make choices, then copyright locks should concern us at *least* as much as influence campaigns. An influence campaign might nudge you to buy a certain brand of phone; but the copyright locks on that phone absolutely determine where you get it serviced, which apps can run on it, and when you have to throw it away rather than fixing it.

Search order and the right to the future tense

Markets are posed as a kind of magic: By discovering otherwise hidden information conveyed by the free choices of consumers, those consumers' local knowledge is integrated into a self-correcting system that makes efficient allocations—more efficient than any computer could calculate. But monopolies are incompatible with that notion. When you only have one app store, the owner of the store—not the consumer—decides on the range of choices. As Boss Tweed once said, "I don't care who does the electing, so long as I get to do the nominating." A monopolized market is an election whose candidates are chosen by the monopolist.

This ballot rigging is made more pernicious by the existence of monopolies over search order. Google's search market share is about 90%. When Google's ranking algorithm puts a result for a popular search term in its top 10, that helps determine the behavior of millions of people. If Google's answer to "Are vaccines dangerous?" is a page that rebuts anti-vax conspiracy theories, then a sizable portion of the public will learn that vaccines are safe. If, on the other hand, Google sends those people to a site affirming the anti-vax conspiracies, a sizable portion of those millions will come away convinced that vaccines are dangerous.

Google's algorithm is often tricked into serving disinformation as a prominent search result. But in these cases, Google isn't persuading people to change their minds; it's just presenting something untrue as fact when the user has no cause to doubt it.

This is true whether the search is for "Are vaccines dangerous?" or "best restaurants near me." Most users will never look past the first page of search results, and when the overwhelming majority of people all use the same search engine, the ranking algorithm deployed by that search engine will determine myriad outcomes (whether to adopt a child, whether to have cancer surgery, where to eat dinner, where to move, where to apply for a job) to a degree that vastly outstrips any behavioral outcomes dictated by algorithmic persuasion techniques.

Many of the questions we ask search engines have no empirically correct answers: "Where should I eat dinner?" is not an objective question. Even questions that do have correct answers ("Are vaccines dangerous?") don't have one empirically superior source for that answer. Many pages affirm the safety of vaccines, so which one goes first? Under conditions of competition, consumers can choose from many search engines and stick with the one whose algorithmic judgment suits them best, but under conditions of monopoly, we all get our answers from the same place.

Google's search dominance isn't a matter of pure merit: The company has leveraged many tactics that would have been prohibited under classical, pre–Ronald Reagan antitrust enforcement standards to attain its dominance. After all, this is a company that has developed two major products: a really good search engine and a pretty good Hotmail clone. Every other major success it's had—Android, YouTube, Google Maps, etc.—has come through an acquisition of a nascent competitor. Many of the company's key divisions, such as the advertising technology of DoubleClick, violate the historical antitrust principle of structural separation, which forbade firms from owning subsidiaries that competed with their customers. Railroads, for example, were barred from owning freight companies that competed with the shippers whose freight they carried.

If we're worried about giant companies subverting markets by stripping consumers of their ability to make free choices, then vigorous antitrust enforcement seems like an excellent remedy. If we'd denied Google the right to effect its many mergers, we would also have probably denied it its total search dominance. Without that dominance, the pet theories, biases, errors (and good judgment, too) of Google search engineers and product managers would not have such an outsized effect on consumer choice.

This goes for many other companies. Amazon, a classic surveillance capitalist, is obviously the dominant tool for searching Amazon—though many people find their way to Amazon through Google searches and Facebook posts—and obviously, Amazon controls Amazon search. That means that Amazon's own self-serving editorial choices—like promoting its own house brands over rival goods from its sellers as well as its own pet theories, biases, and errors—determine much of what we buy on Amazon. And since Amazon is the dominant e-commerce retailer outside of China and since it attained that dominance by buying up both large rivals and nascent competitors in defiance of historical antitrust rules, we can blame the monopoly for stripping consumers of their right to the future tense and the ability to shape markets by making informed choices.

Not every monopolist is a surveillance capitalist, but that doesn't mean they're not able to shape consumer choices in wide-ranging ways. Zuboff lauds Apple for its App Store and iTunes Store, insisting that adding price tags to the features on its platforms has been the secret to resisting surveillance and thus creating markets. But Apple is the only retailer allowed to sell on its platforms, and it's the second-largest mobile device vendor in the world. The independent software vendors that sell through Apple's marketplace accuse the company of the same surveillance sins as Amazon and other big retailers: spying on its customers to find lucrative new products to launch, effectively using independent software vendors as free-market researchers, then forcing them out of any markets they discover.

Because of its use of copyright locks, Apple's mobile customers are not legally allowed to switch to a rival retailer for its apps if they want to do so on an iPhone. Apple, obviously, is the only entity that gets to decide how it ranks the results of search queries in its stores. These decisions ensure that some apps are often installed (because they appear on page one) and others are never installed (because they appear on page one million). Apple's search-ranking design decisions have a vastly more significant effect on consumer behaviors than influence campaigns delivered by surveillance capitalism's ad-serving bots.

Monopolists can afford sleeping pills for watchdogs

Only the most extreme market ideologues think that markets can self-regulate without state oversight. Markets need watchdogs—regulators, lawmakers, and other elements of democratic control—to keep them honest. When these watchdogs sleep on the job, then markets cease to aggregate consumer choices because those choices are constrained by illegitimate and deceptive activities that companies are able to get away with because no one is holding them to account.

But this kind of regulatory capture doesn't come cheap. In competitive sectors, where rivals are constantly eroding one another's margins, individual firms lack the surplus capital to effectively lobby for laws and regulations that serve their ends.

Many of the harms of surveillance capitalism are the result of weak or nonexistent regulation. Those regulatory vacuums spring from the power of monopolists to resist stronger regulation and to tailor what regulation exists to permit their existing businesses.

Here's an example: When firms over-collect and over-retain our data, they are at increased risk of suffering a breach—you can't leak data you never collected, and once you delete all copies of that data, you can no longer leak it. For more than a decade, we've lived through an endless parade of ever-worsening data breaches, each one uniquely horrible in the scale of data breached and the sensitivity of that data.

But still, firms continue to over-collect and over-retain our data for three reasons:

1. **They are locked in the aforementioned limbic arms race with our capacity to shore up our attentional defense systems to resist their new persuasion techniques.** They're also locked in an arms race with their competitors to find new ways to target people for sales pitches. As soon as they discover a soft spot in our attentional defenses (a counterintuitive, unobvious way to target potential refrigerator buyers), the public begins to wise up to the tactic, and their competitors leap on it, hastening the day in which all potential refrigerator buyers have been inured to the pitch.

2. **They believe the surveillance capitalism story.** Data is cheap to aggregate and store, and both proponents and opponents of surveillance capitalism have assured managers and product designers that if you collect enough data, you will be able to perform sorcerous acts of mind control, thus supercharging your sales. Even if you never figure out how

to profit from the data, someone else will eventually offer to buy it from you to give it a try. This is the hallmark of all economic bubbles: acquiring an asset on the assumption that someone else will buy it from you for more than you paid for it, often to sell to someone else at an even greater price.

3. **The penalties for leaking data are negligible.** Most countries limit these penalties to actual damages, meaning that consumers who've had their data breached have to show actual monetary harms to get a reward. In 2014, Home Depot disclosed that it had lost credit-card data for 53 million of its customers, but it settled the matter by paying those customers about $0.34 each—and a third of that $0.34 wasn't even paid in cash. It took the form of a credit to procure a largely ineffectual credit-monitoring service.

But the harms from breaches are much more extensive than these actual-damages rules capture. Identity thieves and fraudsters are wily and endlessly inventive. All the vast breaches of our century are being continuously recombined, the data sets merged and mined for new ways to victimize the people whose data was present in them. Any reasonable, evidence-based theory of deterrence and compensation for breaches would not confine damages to actual damages but rather would allow users to claim these future harms.

However, even the most ambitious privacy rules, such as the EU General Data Protection Regulation, fall far short of capturing the negative externalities of the platforms' negligent over-collection and over-retention, and what penalties they do provide are not aggressively pursued by regulators.

This tolerance of—or indifference to—data over-collection and over-retention can be ascribed in part to the sheer lobbying muscle of the platforms. They are so profitable that they can handily afford to divert gigantic sums to fight any real change—that is, change that would force them to internalize the costs of their surveillance activities.

And then there's state surveillance, which the surveillance capitalism story dismisses as a relic of another era when the big worry was being jailed for your dissident speech, not having your free will stripped away with machine learning.

But state surveillance and private surveillance are intimately related. As we saw when Apple was conscripted by the Chinese government as a vital collaborator in state surveillance, the only really affordable and tractable way to conduct mass surveillance on the scale practiced by modern states—both "free" and autocratic states—is to suborn commercial services.

Whether it's Google being used as a location tracking tool by local law enforcement across the United States or the use of social media tracking by the Department of Homeland Security to build dossiers on participants in protests against Immigration and Customs Enforcement's family separation practices, any hard limits on surveillance capitalism would hamstring the state's own surveillance capability. Without Palantir, Amazon, Google, and other major tech contractors, U.S. cops would not be able to spy on Black people, ICE would not be able to manage the caging of children at the U.S. border, and state welfare systems would not be able to purge their rolls by dressing up cruelty as empiricism and claiming that poor and vulnerable people are ineligible for assistance. At least some of the states' unwillingness to take meaningful action to curb surveillance should be attributed to this symbiotic relationship. There is no mass state surveillance without mass commercial surveillance.

Monopolism is key to the project of mass state surveillance. It's true that smaller tech firms are apt to be less well-defended than Big Tech, whose security experts are drawn from the tops of their field and who are given enormous resources to secure and monitor their systems against intruders. But smaller firms also have less to protect: fewer users whose data is more fragmented across more systems and have to be suborned one at a time by state actors.

A concentrated tech sector that works with authorities is a much more powerful ally in the project of mass state surveillance than a fragmented one composed of smaller actors. The U.S. tech sector is small enough that all of its top executives fit around a single boardroom table in Trump Tower in 2017, shortly after Trump's inauguration. Most of its biggest players bid to win JEDI, the Pentagon's $10 billion Joint Enterprise Defense Infrastructure cloud contract. Like other highly concentrated industries, Big Tech rotates its key employees in and out of government service, sending them to serve in the Department of Defense and the White House, then hiring ex-Pentagon and ex-DOD top staffers and officers to work in their own government relations departments.

They can even make a good case for doing this: After all, when there are only four or five big companies in an industry, everyone qualified to regulate those companies has served as an executive in at least a couple of them—because, likewise, when there are only five companies in an industry, everyone qualified for a senior role at any of them is by definition working at one of the other ones.

Industries that are competitive are fragmented—composed of companies that are at each other's throats all the time and eroding one another's margins in bids to steal their best customers. This leaves them with much more limited capital to use to lobby for favorable rules and a much harder job of getting everyone to agree to pool their resources to benefit the industry as a whole.

Surveillance combined with machine learning is supposed to be an existential crisis, a species-defining moment at which our free will is just a few more advances in the field from being stripped away. I am skeptical of this claim, but I *do* think that tech poses an existential threat to our society and possibly our species.

But that threat grows out of monopoly.

One of the consequences of tech's regulatory capture is that it can shift liability for poor security decisions onto its customers and the wider society. It is absolutely normal in tech for companies to obfuscate the workings of their products, to make them deliberately hard to understand, and to threaten security researchers who seek to independently audit those products.

IT is the only field in which this is practiced: No one builds a bridge or a hospital and keeps the composition of the steel or the equations used to calculate load stresses a secret. It is a frankly bizarre practice that leads, time and again, to grotesque security defects on farcical scales, with whole classes of devices being revealed as vulnerable long after they are deployed in the field and put into sensitive places.

The monopoly power that keeps any meaningful consequences for breaches at bay means that tech companies continue to build terrible products that are insecure by design and that end up integrated into our lives, in possession of our data, and connected to our physical world. For years, Boeing has struggled with the aftermath of a series of bad technology decisions that made its 737 fleet a global pariah, a rare instance in which bad tech decisions have been seriously punished in the market.

These bad security decisions are compounded yet again by the use of copyright locks to enforce business-model decisions against consumers. Recall that these locks have become the go-to means for shaping consumer behavior, making it technically impossible to use third-party ink, insulin, apps, or service depots in connection with your lawfully acquired property.

Recall also that these copyright locks are backstopped by legislation (such as Section 1201 of the DMCA or Article 6 of the 2001 EU Copyright Directive) that ban tampering with ("circumventing") them, and these statutes have been used to threaten security researchers who make disclosures about vulnerabilities without permission from manufacturers.

This amounts to a manufacturer's veto over safety warnings and criticism. While this is far from the legislative intent of the DMCA and its sister statutes around the world, Congress has not intervened to clarify the statute nor will it because to do so would run counter to the interests of powerful, large firms whose lobbying muscle is unstoppable.

Copyright locks are a double whammy: They create bad security decisions that can't be freely investigated or discussed. If markets are supposed to be machines for aggregating information (and if surveillance capitalism's notional mind-control rays are what make it a "rogue capitalism" because it denies consumers the power to make decisions), then a program of legally enforced ignorance of the risks of products makes monopolism even more of a "rogue capitalism" than surveillance capitalism's influence campaigns.

And unlike mind-control rays, enforced silence over security is an immediate, documented problem, and it *does* constitute an existential threat to our civilization and possibly our species. The proliferation of insecure devices—especially devices that spy on us and especially when those devices also can manipulate the physical world by, say, steering your car or flipping a breaker at a power station—is a kind of technology debt.

In software design, "technology debt" refers to old, baked-in decisions that turn out to be bad ones in hindsight. Perhaps a long-ago developer decided to incorporate a networking protocol made by a vendor that has since stopped supporting it. But everything in the product still relies on that superannuated protocol, and so, with each revision, the product team has to work around this obsolete core, adding compatibility layers, surrounding it with security checks that try to shore up its defenses, and so on. These Band-Aid measures compound the debt because every subsequent revision has to make allowances for *them*, too, like interest mounting on a predatory subprime loan. And like a subprime loan, the interest mounts faster than you can hope to pay it off: The product team has to put so much energy into maintaining this complex, brittle system that they don't have any time left over to refactor the product from the ground up and "pay off the debt" once and for all.

Typically, technology debt results in a technological bankruptcy: The product gets so brittle and unsustainable that it fails catastrophically. Think of the antiquated COBOL-based banking and accounting systems that fell over at the start of the pandemic emergency when confronted with surges of unemployment claims. Sometimes that ends the product; sometimes it takes the company down with it. Being caught in the default of a technology debt is scary and traumatic, just like losing your house due to bankruptcy is scary and traumatic.

But the technology debt created by copyright locks isn't individual debt; it's systemic. Everyone in the world is exposed to this over-leverage, as was the case with the 2008 financial crisis. When that debt comes due—when we face a cascade of security breaches that threaten global shipping and logistics, the food supply, pharmaceutical production pipelines, emergency communications, and other critical systems that are accumulating technology debt in part due to the presence of deliberately insecure and deliberately unauditable copyright locks—it will indeed pose an existential risk.

Privacy and monopoly

Many tech companies are gripped by an orthodoxy that holds that if they just gather enough data on enough of our activities, everything else is possible—the mind control and endless profits. This is an unfalsifiable hypothesis: If data gives a tech company even a tiny improvement in behavior prediction and modification, the company declares that it has taken the first step toward global domination with no end in sight. If a company *fails* to attain any improvements from gathering and analyzing data, it declares success to be just around the corner, attainable once more data is in hand.

Surveillance tech is far from the first industry to embrace a nonsensical, self-serving belief that harms the rest of the world, and it is not the first industry to profit handsomely from such a delusion. Long before hedge-fund managers were claiming (falsely) that they could beat the S&P 500, there were plenty of other "respectable" industries that have been revealed as quacks in hindsight. From the makers of radium suppositories (a real thing!) to the cruel sociopaths who claimed they could "cure" gay people, history is littered with the formerly respectable titans of discredited industries.

This is not to say that there's nothing wrong with Big Tech and its ideological addiction to data. While surveillance's benefits are mostly overstated, its harms are, if anything, *understated.*

There's real irony here. The belief in surveillance capitalism as a "rogue capitalism" is driven by the belief that markets wouldn't tolerate firms that are gripped by false beliefs. An oil company that has false beliefs about where the oil is will eventually go broke digging dry wells after all.

But monopolists get to do terrible things for a long time before they pay the price. Think of how concentration in the finance sector allowed the subprime crisis to fester as bond-rating agencies, regulators, investors, and critics all fell under the sway of a false belief that complex mathematics could construct "fully hedged" debt instruments that could not possibly default. A small bank that engaged in this kind of malfeasance would simply go broke rather than outrunning the inevitable crisis, perhaps growing so big that it averted it altogether. But large banks were able to continue to attract investors, and when they finally *did* come a-cropper, the world's governments bailed them out. The worst offenders of the subprime crisis are bigger than they were in 2008, bringing home more profits and paying their execs even larger sums.

Big Tech is able to practice surveillance not just because it is tech but because it is *big*. The reason every web publisher embeds a Facebook "Like" button is that Facebook dominates the internet's social media referrals—and every one of those "Like" buttons spies on everyone who lands on a page that contains them (see also: Google Analytics embeds, Twitter buttons, etc.).

The reason the world's governments have been slow to create meaningful penalties for privacy breaches is that Big Tech's concentration produces huge profits that can be used to lobby against those penalties—and Big Tech's concentration means that the companies involved are able to arrive at a unified negotiating position that supercharges the lobbying.

The reason that the smartest engineers in the world want to work for Big Tech is that Big Tech commands the lion's share of tech industry jobs.

The reason people who are aghast at Facebook's and Google's and Amazon's data-handling practices continue to use these services is that all their friends are on Facebook; Google dominates search; and Amazon has put all the local merchants out of business.

Competitive markets would weaken the companies' lobbying muscle by reducing their profits and pitting them against each other in regulatory forums. It would give customers other places to go to get their online services. It would make the companies small enough to regulate and pave the way to meaningful penalties for breaches. It would let engineers with ideas that challenged the surveillance orthodoxy raise capital to compete with the incumbents. It would give web publishers multiple ways to reach audiences and make the case against Facebook and Google and Twitter embeds.

In other words, while surveillance doesn't cause monopolies, monopolies certainly abet surveillance.

Ronald Reagan, pioneer of tech monopolism

Technology exceptionalism is a sin, whether it's practiced by technology's blind proponents or by its critics. Both of these camps are prone to explaining away monopolistic concentration by citing some special characteristic of the tech industry, like network effects or first-mover advantage. The only real difference between these two groups is that the tech apologists say monopoly is inevitable so we should just let tech get away with its abuses while competition regulators in the United States and the EU say monopoly is inevitable so we should punish tech for its abuses but not try to break up the monopolies.

To understand how tech became so monopolistic, it's useful to look at the dawn of the consumer tech industry: 1979, the year the Apple II Plus launched and became the first successful home computer. That also happens to be the year that Ronald Reagan hit the campaign trail for the 1980 presidential race—a race he won, leading to a radical shift in the way that antitrust concerns are handled in America. Reagan's cohort of politicians—including Margaret Thatcher in the United Kingdom, Brian Mulroney in Canada, Helmut Kohl in Germany, and Augusto Pinochet in Chile—went on to enact similar reforms that eventually spread around the world.

Antitrust's story began nearly a century before all that with laws like the Sherman Act, which took aim at monopolists on the grounds that monopolies were bad in and of themselves—squeezing out competitors, creating "diseconomies of scale" (when a company is so big that its constituent parts go awry and it is seemingly helpless to address the problems), and capturing their regulators to such a degree that they can get away with a host of evils.

Then came a fabulist named Robert Bork, a former solicitor general who Reagan appointed to the powerful U.S. Court of Appeals for the D.C. Circuit and who had created an alternate legislative history of the Sherman Act and its successors out of whole cloth. Bork insisted that these statutes were never targeted at monopolies (despite a wealth of evidence to the contrary, including the transcribed speeches of the acts' authors) but, rather, that they were intended to prevent "consumer harm"—in the form of higher prices.

Bork was a crank, but he was a crank with a theory that rich people really liked. Monopolies are a great way to make rich people richer by allowing them to receive "monopoly rents" (that is, bigger profits) and capture regulators, leading to a weaker, more favorable regulatory environment with fewer protections for customers, suppliers, the environment, and workers.

Bork's theories were especially palatable to the same power brokers who backed Reagan, and Reagan's Department of Justice and other agencies began to incorporate Bork's antitrust doctrine into their enforcement decisions (Reagan even put Bork up for a Supreme Court seat, but Bork flunked the Senate confirmation hearing so badly that, 40 years later, D.C. insiders use the term "borked" to refer to any catastrophically bad political performance).

Little by little, Bork's theories entered the mainstream, and their backers began to infiltrate the legal education field, even putting on junkets where members of the judiciary were treated to lavish meals, fun outdoor activities, and seminars where they were indoctrinated into the consumer harm theory of antitrust. The more Bork's theories took hold, the more money the monopolists were making—and the more surplus capital they had at their disposal to lobby for even more Borkian antitrust influence campaigns.

The history of Bork's antitrust theories is a really good example of the kind of covertly engineered shifts in public opinion that Zuboff warns us against, where fringe ideas become mainstream orthodoxy. But Bork didn't change the world overnight. He played a very long game, for over a generation, and he had a tailwind because the same forces that backed oligarchic antitrust theories also backed many other oligarchic shifts in public opinion. For example, the idea that taxation is theft, that wealth is a sign of virtue, and so on—all of these theories meshed to form a coherent ideology that elevated inequality to a virtue.

Today, many fear that machine learning allows surveillance capitalism to sell "Bork-as-a-Service," at internet speeds, so that you can contract a machine-learning company to engineer *rapid* shifts in public sentiment without needing the capital to sustain a multipronged, multigenerational project working at the local, state, national, and global levels in business, law, and philosophy. I do not believe that such a project is plausible, though I agree that this is basically what the platforms claim to be selling. They're just lying about it. Big Tech lies all the time, *including* in their sales literature.

The idea that tech forms "natural monopolies" (monopolies that are the inevitable result of the realities of an industry, such as the monopolies that accrue the first company to run long-haul phone lines or rail lines) is belied by tech's own history: In the absence of anti-competitive tactics, Google was able to unseat AltaVista and Yahoo; Facebook was able to head off Myspace. There are some advantages to gathering mountains of data, but those mountains of data also have disadvantages: liability (from leaking), diminishing returns (from old data), and institutional inertia (big companies, like science, progress one funeral at a time).

Indeed, the birth of the web saw a mass-extinction event for the existing giant, wildly profitable proprietary technologies that had capital, network effects, and walls and moats surrounding their businesses. The web showed that when a new industry is built around a protocol, rather than a product, the combined might of everyone who uses the protocol to reach their customers or users or communities outweighs even the most massive products. CompuServe, AOL, MSN, and a host of other proprietary walled gardens learned this lesson the hard way: Each believed it could stay separate from the web, offering "curation" and a guarantee of consistency and quality instead of the chaos of an open system. Each was wrong and ended up being absorbed into the public web.

Yes, tech is heavily monopolized and is now closely associated with industry concentration, but this has more to do with a matter of timing than its intrinsically monopolistic tendencies. Tech was born at the moment that antitrust enforcement was being dismantled, and tech fell into exactly the same pathologies that antitrust was supposed to guard against. To a first approximation, it is reasonable to assume that tech's monopolies are the result of a lack of anti-monopoly action and not the much-touted unique characteristics of tech, such as network effects, first-mover advantage, and so on.

In support of this thesis, I offer the concentration that every *other* industry has undergone over the same period. From professional wrestling to consumer packaged goods to commercial property leasing to banking to sea freight to oil to record labels to newspaper ownership to theme parks, *every* industry has undergone a massive shift toward concentration. There's no obvious network effects or first-mover advantage at play in these industries. However, in every case, these industries attained their concentrated status through tactics that were prohibited before Bork's triumph: merging with major competitors, buying out innovative new market entrants, horizontal and vertical integration, and a suite of anti-competitive tactics that were once illegal but are not any longer.

Again: When you change the laws intended to prevent monopolies and then monopolies form in exactly the way the law was supposed to prevent, it is reasonable to suppose that these facts are related. Tech's concentration can be readily explained without recourse to radical theories of network effects—but only if you're willing to indict unregulated markets as tending toward monopoly. Just as a lifelong smoker can give you a hundred reasons why their smoking didn't cause their cancer ("It was the environmental toxins"), true believers in unregulated markets have a whole suite of unconvincing explanations for monopoly in tech that leave capitalism intact.

Steering with the windshield wipers

It's been 40 years since Bork's project to rehabilitate monopolies achieved liftoff, and that is a generation and a half, which is plenty of time to take a common idea and make it seem outlandish and vice versa. Before the 1940s, affluent Americans dressed their baby boys in pink while baby girls wore blue (a "delicate and dainty" color). While gendered colors are obviously totally arbitrary, many still greet this news with amazement and find it hard to imagine a time when pink connoted masculinity.

After 40 years of studiously ignoring antitrust analysis and enforcement, it's not surprising that we've all but forgotten that antitrust exists, that in living memory, growth through mergers and acquisitions were largely prohibited under law, that market-cornering strategies like vertical integration could land a company in court.

Antitrust is a market society's steering wheel, the control of first resort to keep would-be masters of the universe in their lanes. But Bork and his cohort ripped out our steering wheel 40 years ago. The car is still barreling along, and so we're yanking as hard as we can on all the *other* controls in the car as well as desperately flapping the doors and rolling the windows up and down in the hopes that one of these other controls can be repurposed to let us choose where we're heading before we careen off a cliff.

It's like a 1960s science-fiction plot come to life: People stuck in a "generation ship," plying its way across the stars, a ship once piloted by their ancestors; and now, after a great cataclysm, the ship's crew have forgotten that they're in a ship at all and no longer remember where the control room is. Adrift, the ship is racing toward its extinction, and unless we can seize the controls and execute emergency course correction, we're all headed for a fiery death in the heart of a sun.

Surveillance still matters

None of this is to minimize the problems with surveillance. Surveillance matters, and Big Tech's use of surveillance *is* an existential risk to our species, but that's not because surveillance and machine learning rob us of our free will.

Surveillance has become *much* more efficient thanks to Big Tech. In 1989, the Stasi—the East German secret police—had the whole country under surveillance, a massive undertaking that recruited one out of every 60 people to serve as an informant or intelligence operative.

Today, we know that the NSA is spying on a significant fraction of the entire world's population, and its ratio of surveillance operatives to the surveilled is more like 1:10,000 (that's probably on the low side since it assumes that every American with top-secret clearance is working for the NSA on this project—we don't know how many of those cleared people are involved in NSA spying, but it's definitely not all of them).

How did the ratio of surveillable citizens expand from 1:60 to 1:10,000 in less than 30 years? It's thanks to Big Tech. Our devices and services gather most of the data that the NSA mines for its surveillance project. We pay for these devices and the services they connect to, and then we painstakingly perform the data-entry tasks associated with logging facts about our lives, opinions, and preferences. This mass surveillance project has been largely useless for fighting terrorism: The NSA can only point to a single minor success story in which it used its data collection program to foil an attempt by a U.S. resident to wire a few thousand dollars to an overseas terror group. It's ineffective for much the same reason that commercial surveillance projects are largely ineffective at targeting advertising: The people who want to commit acts of terror, like people who want to buy a refrigerator, are extremely rare. If you're trying to detect a phenomenon whose base rate is one in a million with an instrument whose accuracy is only 99%, then every true positive will come at the cost of 9,999 false positives.

Let me explain that again: If one in a million people is a terrorist, then there will only be about one terrorist in a random sample of one million people. If your test for detecting terrorists is 99% accurate, it will identify 10,000 terrorists in your million-person sample (1% of one million is 10,000). For every true positive, you'll get 9,999 false positives.

In reality, the accuracy of algorithmic terrorism detection falls far short of the 99% mark, as does refrigerator ad targeting.

The difference is that being falsely accused of wanting to buy a fridge is a minor nuisance while being falsely accused of planning a terror attack can destroy your life and the lives of everyone you love.

Mass state surveillance is only feasible because of surveillance capitalism and its extremely low-yield ad-targeting systems, which require a constant feed of personal data to remain barely viable. Surveillance capitalism's primary failure mode is mistargeted ads while mass state surveillance's primary failure mode is grotesque human rights abuses, tending toward totalitarianism.

State surveillance is no mere parasite on Big Tech, sucking up its data and giving nothing in return. In truth, the two are symbiotes: Big Tech sucks up our data for spy agencies, and spy agencies ensure that governments don't limit Big Tech's activities so severely that it would no longer serve the spy agencies' needs. There is no firm distinction between state surveillance and surveillance capitalism; they are dependent on one another.

To see this at work today, look no further than Amazon's home surveillance device, the Ring doorbell, and its associated app, Neighbors. Ring—a product that Amazon acquired and did not develop in house—makes a camera-enabled doorbell that streams footage from your front door to your mobile device. The Neighbors app allows you to form a neighborhood-wide surveillance grid with your fellow Ring owners through which you can share clips of "suspicious characters." If you're thinking that this sounds like a recipe for letting curtain-twitching racists supercharge their suspicions of people with brown skin who walk down their blocks, you're right. Ring has become a de facto, off-the-books arm of the police without any of the pesky oversight or rules.

In mid-2019, a series of public records requests revealed that Amazon had struck confidential deals with more than 400 local law enforcement agencies through which the agencies would promote Ring and Neighbors and in exchange get access to footage from Ring cameras. In theory, cops would need to request this footage through Amazon (and internal documents reveal that Amazon devotes substantial resources to coaching cops on how to spin a convincing story when doing so), but in practice, when a Ring customer turns down a police request, Amazon only requires the agency to formally request the footage from the company, which it will then produce.

Ring and law enforcement have found many ways to intertwine their activities. Ring strikes secret deals to acquire real-time access to 911 dispatch and then streams alarming crime reports to Neighbors users, which serve as convincers for anyone who's contemplating a surveillance doorbell but isn't sure whether their neighborhood is dangerous enough to warrant it.

The more the cops buzz-market the surveillance capitalist Ring, the more surveillance capability the state gets. Cops who rely on private entities for law-enforcement roles then brief against any controls on the deployment of that technology while the companies return the favor by lobbying against rules requiring public oversight of police surveillance technology. The more the cops rely on Ring and Neighbors, the harder it will be to pass laws to curb them. The fewer laws there are against them, the more the cops will rely on them.

Dignity and sanctuary

But even if we could exercise democratic control over our states and force them to stop raiding surveillance capitalism's reservoirs of behavioral data, surveillance capitalism would still harm us.

This is an area where Zuboff shines. Her chapter on "sanctuary"—the feeling of being unobserved—is a beautiful hymn to introspection, calmness, mindfulness, and tranquility.

When you are watched, something changes. Anyone who has ever raised a child knows this. You might look up from your book (or more realistically, from your phone) and catch your child in a moment of profound realization and growth, a moment where they are learning something that is right at the edge of their abilities, requiring their entire ferocious concentration. For a moment, you're transfixed, watching that rare and beautiful moment of focus playing out before your eyes, and then your child looks up and sees you seeing them, and the moment collapses. To grow, you need to be and expose your authentic self, and in that moment, you are vulnerable like a hermit crab scuttling from one shell to the next. The tender, unprotected tissues you expose in that moment are too delicate to reveal in the presence of another, even someone you trust as implicitly as a child trusts their parent.

In the digital age, our authentic selves are inextricably tied to our digital lives. Your search history is a running ledger of the questions you've pondered. Your location history is a record of the places you've sought out and the experiences you've had there. Your social graph reveals the different facets of your identity, the people you've connected with.

To be observed in these activities is to lose the sanctuary of your authentic self.

There's another way in which surveillance capitalism robs us of our capacity to be our authentic selves: by making us anxious. Surveillance capitalism isn't really a mind-control ray, but you don't need a mind-control ray to make someone anxious. After all, another word for anxiety is agitation, and to make someone experience agitation, you need merely to agitate them. To poke them and prod them and beep at them and buzz at them and bombard them on an intermittent schedule that is just random enough that our limbic systems never quite become inured to it.

Our devices and services are "general purpose" in that they can connect anything or anyone to anything or anyone else and that they can run any program that can be written. This means that the distraction rectangles in our pockets hold our most precious moments with our most beloved people and their most urgent or time-sensitive communications (from "running late can you get the kid?" to "doctor gave me bad news and I need to talk to you RIGHT NOW") as well as ads for refrigerators and recruiting messages from Nazis.

All day and all night, our pockets buzz, shattering our concentration and tearing apart the fragile webs of connection we spin as we think through difficult ideas. If you locked someone in a cell and agitated them like this, we'd call it "sleep deprivation torture," and it would be a war crime under the Geneva Conventions.

Afflicting the afflicted

The effects of surveillance on our ability to be our authentic selves are not equal for all people. Some of us are lucky enough to live in a time and place in which all the most important facts of our lives are widely and roundly socially acceptable and can be publicly displayed without the risk of social consequence.

But for many of us, this is not true. Recall that in living memory, many of the ways of being that we think of as socially acceptable today were once cause for dire social sanction or even imprisonment. If you are 65 years old, you have lived through a time in which people living in "free societies" could be imprisoned or sanctioned for engaging in homosexual activity, for falling in love with a person whose skin was a different color than their own, or for smoking weed.

Today, these activities aren't just decriminalized in much of the world, they're considered normal, and the fallen prohibitions are viewed as shameful, regrettable relics of the past.

How did we get from prohibition to normalization? Through private, personal activity: People who were secretly gay or secret pot-smokers or who secretly loved someone with a different skin color were vulnerable to retaliation if they made their true selves known and were limited in how much they could advocate for their own right to exist in the world and be true to themselves. But because there was a private sphere, these people could form alliances with their friends and loved ones who did not share their disfavored traits by having private conversations in which they came out, disclosing their true selves to the people around them and bringing them to their cause one conversation at a time.

The right to choose the time and manner of these conversations was key to their success. It's one thing to come out to your dad while you're on a fishing trip away from the world and another thing entirely to blurt it out over the Christmas dinner table while your racist Facebook uncle is there to make a scene.

Without a private sphere, there's a chance that none of these changes would have come to pass and that the people who benefited from these changes would have either faced social sanction for coming out to a hostile world or would have never been able to reveal their true selves to the people they love.

The corollary is that, unless you think that our society has attained social perfection—that your grandchildren in 50 years will ask you to tell them the story of how, in 2020, every injustice had been righted and no further change had to be made—then you should expect that right now, at this minute, there are people you love, whose happiness is key to your own, who have a secret in their hearts that stops them from ever being their authentic selves with you. These people are sorrowing and will go to their graves with that secret sorrow in their hearts, and the source of that sorrow will be the falsity of their relationship to you.

A private realm is necessary for human progress.

Any data you collect and retain will eventually leak

The lack of a private life can rob vulnerable people of the chance to be their authentic selves and constrain our actions by depriving us of sanctuary but there is another risk that is borne by everyone, not just people with a secret: crime.

Personally identifying information is of very limited use for the purpose of controlling peoples' minds, but identity theft—really a catchall term for a whole constellation of terrible criminal activities that can destroy your finances, compromise your personal integrity, ruin your reputation, or even expose you to physical danger—thrives on it.

Attackers are not limited to using data from one breached source, either. Multiple services have suffered breaches that exposed names, addresses, phone numbers, passwords, sexual tastes, school grades, work performance, brushes with the criminal justice system, family details, genetic information, fingerprints and other biometrics, reading habits, search histories, literary tastes, pseudonymous identities, and other sensitive information. Attackers can merge data from these different breaches to build up extremely detailed dossiers on random subjects and then use different parts of the data for different criminal purposes.

For example, attackers can use leaked username and password combinations to hijack whole fleets of commercial vehicles that have been fitted with anti-theft GPS trackers and immobilizers or to hijack baby monitors in order to terrorize toddlers with the audio tracks from pornography. Attackers use leaked data to trick phone companies into giving them your phone number, then they intercept SMS-based two-factor authentication codes in order to take over your email, bank account, and/or cryptocurrency wallets.

Attackers are endlessly inventive in the pursuit of creative ways to weaponize leaked data. One common use of leaked data is to penetrate companies in order to access *more* data.

Like spies, online fraudsters are totally dependent on companies over-collecting and over-retaining our data. Spy agencies sometimes pay companies for access to their data or intimidate them into giving it up, but sometimes they work just like criminals do—by sneaking data out of companies' databases.

The over-collection of data has a host of terrible social consequences, from the erosion of our authentic selves to the undermining of social progress, from state surveillance to an epidemic of online crime. Commercial surveillance is also a boon to people running influence campaigns, but that's the least of our troubles.

Critical tech exceptionalism is still tech exceptionalism

Big Tech has long practiced technology exceptionalism: the idea that it should not be subject to the mundane laws and norms of "meatspace." Mottoes like Facebook's "move fast and break things" attracted justifiable scorn of the companies' self-serving rhetoric.

Tech exceptionalism got us all into a lot of trouble, so it's ironic and distressing to see Big Tech's critics committing the same sin.

Big Tech is not a "rogue capitalism" that cannot be cured through the traditional anti-monopoly remedies of trustbusting (forcing companies to divest of competitors they have acquired) and bans on mergers to monopoly and other anti-competitive tactics. Big Tech does not have the power to use machine learning to influence our behavior so thoroughly that markets lose the ability to punish bad actors and reward superior competitors. Big Tech has no rule-writing mind-control ray that necessitates ditching our old toolbox.

The thing is, people have been claiming to have perfected mind-control rays for centuries, and every time, it turned out to be a con—though sometimes the con artists were also conning themselves.

For generations, the advertising industry has been steadily improving its ability to sell advertising services to businesses while only making marginal gains in selling those businesses' products to prospective customers. John Wanamaker's lament that "50% of my advertising budget is wasted, I just don't know which 50%" is a testament to the triumph of *ad executives*, who successfully convinced Wanamaker that only half of the money he spent went to waste.

The tech industry has made enormous improvements in the science of convincing businesses that they're good at advertising while their actual improvements to advertising—as opposed to targeting—have been pretty ho-hum. The vogue for machine learning—and the mystical invocation of "artificial intelligence" as a synonym for straightforward statistical inference techniques—has greatly boosted the efficacy of Big Tech's sales pitch as marketers have exploited potential customers' lack of technical sophistication to get away with breathtaking acts of overpromising and underdelivering.

It's tempting to think that if businesses are willing to pour billions into a venture that the venture must be a good one. Yet there are plenty of times when this rule of thumb has led us astray. For example, it's virtually unheard of for managed investment funds to outperform simple index funds, and investors who put their money into the hands of expert money managers overwhelmingly fare worse than those who entrust their savings to index funds. But managed funds still account for the majority of the money invested in the markets, and they are patronized by some of the richest, most sophisticated investors in the world. Their vote of confidence in an underperforming sector is a parable about the role of luck in wealth accumulation, not a sign that managed funds are a good buy.

The claims of Big Tech's mind-control system are full of tells that the enterprise is a con. For example, the reliance on the "Big Five" personality traits as a primary means of influencing people even though the "Big Five" theory is unsupported by any large-scale, peer-reviewed studies and is mostly the realm of marketing hucksters and pop psych.

Big Tech's promotional materials also claim that their algorithms can accurately perform "sentiment analysis" or detect peoples' moods based on their "microexpressions," but these are marketing claims, not scientific ones. These methods are largely untested by independent scientific experts, and where they have been tested, they've been found sorely wanting. Microexpressions are particularly suspect as the companies that specialize in training people to detect them have been shown to underperform relative to random chance.

Big Tech has been so good at marketing its own supposed superpowers that it's easy to believe that they can market everything else with similar acumen, but it's a mistake to believe the hype. Any statement a company makes about the quality of its products is clearly not impartial. The fact that we distrust all the things that Big Tech says about its data handling, compliance with privacy laws, etc., is only reasonable—but why on Earth would we treat Big Tech's marketing literature as the gospel truth? Big Tech lies about just about *everything*, including how well its machine-learning fueled persuasion systems work.

That skepticism should infuse all of our evaluations of Big Tech and its supposed abilities, including our perusal of its patents. Zuboff vests these patents with enormous significance, pointing out that Google claimed extensive new persuasion capabilities in its patent filings. These claims are doubly suspect: first, because they are so self-serving, and second, because the patent itself is so notoriously an invitation to exaggeration.

Patent applications take the form of a series of claims and range from broad to narrow. A typical patent starts out by claiming that its authors have invented a method or system for doing every conceivable thing that anyone might do, ever, with any tool or device. Then it narrows that claim in successive stages until we get to the actual "invention" that is the true subject of the patent. The hope is that the patent examiner—who is almost certainly overworked and underinformed—will miss the fact that some or all of these claims are ridiculous, or at least suspect, and grant the patent's broader claims. Patents for unpatentable things are still incredibly useful because they can be wielded against competitors who might license that patent or steer clear of its claims rather than endure the lengthy, expensive process of contesting it.

What's more, software patents are routinely granted even though the filer doesn't have any evidence that they can do the thing claimed by the patent. That is, you can patent an "invention" that you haven't actually made and that you don't know how to make.

With these considerations in hand, it becomes obvious that the fact that a Big Tech company has patented what it *says* is an effective mind-control ray is largely irrelevant to whether Big Tech can in fact control our minds.

Big Tech collects our data for many reasons, including the diminishing returns on existing stores of data. But many tech companies also collect data out of a mistaken tech exceptionalist belief in the network effects of data. Network effects occur when each new user in a system increases its value. The classic example is fax machines: A single fax machine is of no use, two fax machines are of limited use, but every new fax machine that's put to use after the first doubles the number of possible fax-to-fax links.

Data mined for predictive systems doesn't necessarily produce these dividends. Think of Netflix: The predictive value of the data mined from a million English-speaking Netflix viewers is hardly improved by the addition of one more user's viewing data. Most of the data Netflix acquires after that first minimum viable sample duplicates existing data and produces only minimal gains. Meanwhile, retraining models with new data gets progressively more expensive as the number of data points increases, and manual tasks like labeling and validating data do not get cheaper at scale.

Businesses pursue fads to the detriment of their profits all the time, especially when the businesses and their investors are not motivated by the prospect of becoming profitable but rather by the prospect of being acquired by a Big Tech giant or by having an IPO. For these firms, ticking faddish boxes like "collects as much data as possible" might realize a bigger return on investment than "collects a business-appropriate quantity of data."

This is another harm of tech exceptionalism: The belief that more data always produces more profits in the form of more insights that can be translated into better mind-control rays drives firms to over-collect and over-retain data beyond all rationality. And since the firms are behaving irrationally, a good number of them will go out of business and become ghost ships whose cargo holds are stuffed full of data that can harm people in myriad ways—but which no one is responsible for any longer. Even if the companies don't go under, the data they collect is maintained behind the minimum viable security—just enough security to keep the company viable while it waits to get bought out by a tech giant, an amount calculated to spend not one penny more than is necessary on protecting data.

How monopolies, not mind control, drive surveillance capitalism: The Snapchat story

For the first decade of its existence, Facebook competed with the social media giants of the day (Myspace, Orkut, etc.) by presenting itself as the pro-privacy alternative. Indeed, Facebook justified its walled garden—which let users bring in data from the web but blocked web services like Google Search from indexing and caching Facebook pages—as a pro-privacy measure that protected users from the surveillance-happy winners of the social media wars like Myspace.

Despite frequent promises that it would never collect or analyze its users' data, Facebook periodically created initiatives that did just that, like the creepy, ham-fisted Beacon tool, which spied on you as you moved around the web and then added your online activities to your public timeline, allowing your friends to monitor your browsing habits. Beacon sparked a user revolt. Every time, Facebook backed off from its surveillance initiative, but not all the way; inevitably, the new

Facebook would be more surveilling than the old Facebook, though not quite as surveilling as the intermediate Facebook following the launch of the new product or service.

The pace at which Facebook ramped up its surveillance efforts seems to have been set by Facebook's competitive landscape. The more competitors Facebook had, the better it behaved. Every time a major competitor foundered, Facebook's behavior got markedly worse.

All the while, Facebook was prodigiously acquiring companies, including a company called Onavo. Nominally, Onavo made a battery-monitoring mobile app. But the permissions that Onavo required were so expansive that the app was able to gather fine-grained telemetry on everything users did with their phones, including which apps they used and how they were using them.

Through Onavo, Facebook discovered that it was losing market share to Snapchat, an app that—like Facebook a decade before—billed itself as the pro-privacy alternative to the status quo. Through Onavo, Facebook was able to mine data from the devices of Snapchat users, including both current and former Snapchat users. This spurred Facebook to acquire Instagram—some features of which competed with Snapchat—and then allowed Facebook to fine-tune Instagram's features and sales pitch to erode Snapchat's gains and ensure that Facebook would not have to face the kinds of competitive pressures it had earlier inflicted on Myspace and Orkut.

The story of how Facebook crushed Snapchat reveals the relationship between monopoly and surveillance capitalism.

Facebook combined surveillance with lax antitrust enforcement to spot the competitive threat of Snapchat on its horizon and then take decisive action against it. Facebook's surveillance capitalism let it avert competitive pressure with anti-competitive tactics. Facebook users still want privacy—Facebook hasn't used surveillance to brainwash them out of it—but they can't get it because Facebook's surveillance lets it destroy any hope of a rival service emerging that competes on privacy features.

A monopoly over your friends

A decentralization movement has tried to erode the dominance of Facebook and other Big Tech companies by fielding "indieweb" alternatives—Mastodon as a Twitter alternative, Diaspora as a Facebook alternative, etc.—but these efforts have failed to attain any kind of liftoff.

Fundamentally, each of these services is hamstrung by the same problem: Every potential user for a Facebook or Twitter alternative has to convince all their friends to follow them to a decentralized web alternative in order to continue to realize the benefit of social media. For many of us, the only reason to have a Facebook account is that our friends have Facebook accounts, and the reason they have Facebook accounts is that *we* have Facebook accounts.

All of this has conspired to make Facebook—and other dominant platforms—into "kill zones" that investors will not fund new entrants for.

And yet, all of today's tech giants came into existence despite the entrenched advantage of the companies that came before them. To understand how that happened, you have to understand both interoperability and adversarial interoperability.

"Interoperability" is the ability of two technologies to work with one another: Anyone can make an LP that will play on any record player, anyone can make a filter you can install in your stove's extractor fan, anyone can make gasoline for your car, anyone can make a USB phone charger that fits in your car's cigarette lighter receptacle, anyone can make a light bulb that works in your light socket, anyone can make bread that will toast in your toaster.

Interoperability is often a source of innovation and consumer benefit: Apple made the first commercially successful PC, but millions of independent software vendors made interoperable programs that ran on the Apple II Plus. The simple analog antenna inputs on the back of TVs first allowed cable operators to connect directly to TVs, then they allowed game console companies and then personal computer companies to use standard as displays. Standard RJ-11 telephone jacks allowed for the production of phones from a variety of vendors in a variety of forms, from the free football-shaped phone that came with a *Sports Illustrated* subscription to business phones with speakers, hold functions, and so on and then answering machines and finally modems, paving the way for the internet revolution.

"Interoperability" is often used interchangeably with "standardization," which is the process manufacturers and other stakeholders hammer out a set of agreed-upon rules for implementing a technology, such as the electrical plug on your wall, the CAN bus used by your car's computer systems, or the HTML instructions that your browser interprets.

But interoperability doesn't require standardization—indeed, standardization often proceeds from the chaos of ad hoc interoperability measures. The inventor of the cigarette-lighter USB charger didn't need to get permission from car manufacturers or even the manufacturers of the dashboard lighter subcomponent. The automakers didn't take any countermeasures to prevent the use of these aftermarket accessories by their customers, but they also didn't do anything to make life easier for the chargers' manufacturers. This is a kind of "neutral interoperability."

Beyond neutral interoperability, there is "adversarial interoperability." That's when a manufacturer makes a product that interoperates with another manufacturer's product *despite the second manufacturer's objections* and *even if that means bypassing a security system designed to prevent interoperability*.

Probably the most familiar form of adversarial interoperability is third-party printer ink. Printer manufacturers claim that they sell printers below cost and that the only way they can recoup the losses they incur is by charging high markups on ink. To prevent the owners of printers from buying ink elsewhere, the printer companies deploy a suite of anti-customer security systems that detect and reject both refilled and third-party cartridges.

Owners of printers take the position that HP and Epson and Brother are not charities and that customers for their wares have no obligation to help them survive, and so if the companies choose to sell their products at a loss, that's their foolish choice and their consequences to live with. Likewise, competitors who make ink or refill kits observe that they don't owe printer companies anything, and their erosion of printer companies' margins are the printer companies' problems, not their competitors'. After all, the printer companies shed no tears when they drive a refiller out of business, so why should the refillers concern themselves with the economic fortunes of the printer companies?

Adversarial interoperability has played an outsized role in the history of the tech industry: from the founding of the "alt.*" Usenet hierarchy (which was started against the wishes of Usenet's maintainers and which grew to be bigger than all of Usenet combined) to the browser wars (when Netscape and Microsoft devoted massive engineering efforts to making their browsers incompatible with the other's special commands and peccadilloes) to Facebook (whose success was built in part by helping its new users stay in touch with friends they'd left behind on Myspace because Facebook supplied them with a tool that scraped waiting messages from Myspace and imported them into Facebook, effectively creating an Facebook-based Myspace reader).

Today, incumbency is seen as an unassailable advantage. Facebook is where all of your friends are, so no one can start a Facebook competitor. But adversarial compatibility reverses the competitive advantage: If you were allowed to compete with Facebook by providing a tool that imported all your users' waiting Facebook messages into an environment that competed on lines that Facebook couldn't cross, like eliminating surveillance and ads, then Facebook would be at a huge disadvantage. It would have assembled all possible ex-Facebook users into a single, easy-to-find service; it would have educated them on how a Facebook-like service worked and what its potential benefits were; and it would have provided an easy means for disgruntled Facebook users to tell their friends where they might expect better treatment.

Adversarial interoperability was once the norm and a key contributor to the dynamic, vibrant tech scene, but now it is stuck behind a thicket of laws and regulations that add legal risks to the tried-and-true tactics of adversarial interoperability. New rules and new interpretations of existing rules mean that a would-be adversarial interoperator needs to steer clear of claims under copyright, terms of service, trade secrecy, tortious interference, and patent.

In the absence of a competitive market, lawmakers have resorted to assigning expensive, state-like duties to Big Tech firms, such as automatically filtering user contributions for copyright infringement or terrorist and extremist content or detecting and preventing harassment in real time or controlling access to sexual material.

These measures put a floor under how small we can make Big Tech because only the very largest companies can afford the humans and automated filters needed to perform these duties.

But that's not the only way in which making platforms responsible for policing their users undermines competition.

A platform that is expected to police its users' conduct must prevent many vital adversarial interoperability techniques lest these subvert its policing measures. For example, if someone using a Twitter replacement like Mastodon is able to push messages into Twitter and read messages out of Twitter, they could avoid being caught by automated systems that detect and prevent harassment (such as systems that use the timing of messages or IP-based rules to make guesses about whether someone is a harasser).

To the extent that we are willing to let Big Tech police itself—rather than making Big Tech small enough that users can leave bad platforms for better ones and small enough that a regulation that simply puts a platform out of business will not destroy billions of users' access to their communities and data—we build the case that Big Tech should be able to block its competitors and make it easier for Big Tech to demand legal enforcement tools to ban and punish attempts at adversarial interoperability.

Ultimately, we can try to fix Big Tech by making it responsible for bad acts by its users, or we can try to fix the internet by cutting Big Tech down to size. But we can't do both. To replace today's giant products with pluralistic protocols, we need to clear the legal thicket that prevents adversarial interoperability so that tomorrow's nimble, personal, small-scale products can federate themselves with giants like Facebook, allowing the users who've left to continue to communicate with users who haven't left yet, reaching tendrils over Facebook's garden wall that Facebook's trapped users can use to scale the walls and escape to the global, open web.

Fake news is an epistemological crisis

Tech is not the only industry that has undergone massive concentration since the Reagan era. Virtually every major industry—from oil to newspapers to meatpacking to sea freight to eyewear to online pornography—has become a clubby oligarchy that just a few players dominate.

At the same time, every industry has become something of a tech industry as general-purpose computers and general-purpose networks and the promise of efficiencies through data-driven analysis infuse every device, process, and firm with tech.

This phenomenon of industrial concentration is part of a wider story about wealth concentration overall as a smaller and smaller number of people own more and more of our world. This concentration of both wealth and industries means that our political outcomes are increasingly beholden to the parochial interests of the people and companies with all the money.

That means that whenever a regulator asks a question with an obvious, empirical answer ("Are humans causing climate change?" or "Should we let companies conduct commercial mass surveillance?" or "Does society benefit from allowing network neutrality violations?"), the answer that comes out is only correct if that correctness meets with the approval of rich people and the industries that made them so wealthy.

Rich people have always played an outsized role in politics and more so since the Supreme Court's *Citizens United* decision eliminated key controls over political spending. Widening inequality and wealth concentration means that the very richest people are now a lot richer and can afford to spend a lot more money on political projects than ever before. Think of the Koch brothers or George Soros or Bill Gates.

But the policy distortions of rich individuals pale in comparison to the policy distortions that concentrated industries are capable of. The companies in highly concentrated industries are much more profitable than companies in competitive industries—no competition means not having to reduce prices or improve quality to win customers—leaving them with bigger capital surpluses to spend on lobbying.

Concentrated industries also find it easier to collaborate on policy objectives than competitive ones. When all the top execs from your industry can fit around a single boardroom table, they often do. And *when* they do, they can forge a consensus position on regulation.

Rising through the ranks in a concentrated industry generally means working at two or three of the big companies. When there are only relatively few companies in a given industry, each company has a more ossified executive rank, leaving ambitious execs with fewer paths to higher positions unless they are recruited to a rival. This means that the top execs in concentrated industries are likely to have been colleagues at some point and socialize in the same circles—connected through social ties or, say, serving as trustees for each others' estates. These tight social bonds foster a collegial, rather than competitive, attitude.

Highly concentrated industries also present a regulatory conundrum. When an industry is dominated by just four or five companies, the only people who are likely to truly understand the industry's practices are its veteran executives. This means that top regulators are often former execs of the companies they are supposed to be regulating. These turns in government are often tacitly understood to be leaves of absence from industry, with former employers welcoming their erstwhile watchdogs back into their executive ranks once their terms have expired.

All this is to say that the tight social bonds, small number of firms, and regulatory capture of concentrated industries give the companies that comprise them the power to dictate many, if not all, of the regulations that bind them.

This is increasingly obvious. Whether it's payday lenders winning the right to practice predatory lending or Apple winning the right to decide who can fix your phone or Google and Facebook winning the right to breach your private data without suffering meaningful consequences or victories for pipeline companies or impunity for opioid manufacturers or massive tax subsidies for incredibly profitable dominant businesses, it's increasingly apparent that many of our official, evidence-based truth-seeking processes are, in fact, auctions for sale to the highest bidder.

It's really impossible to overstate what a terrifying prospect this is. We live in an incredibly high-tech society, and none of us could acquire the expertise to evaluate every technological proposition that stands between us and our untimely, horrible deaths. You might devote your life to acquiring the media literacy to distinguish good scientific journals from corrupt pay-for-play lookalikes and the statistical literacy to evaluate the quality of the analysis in the journals as well as the microbiology and epidemiology knowledge to determine whether you can trust claims about the safety of vaccines—but that would still leave you unqualified to judge whether the wiring in your home will give you a lethal shock *and* whether your car's brakes' software will cause them to fail unpredictably *and* whether the hygiene standards at your butcher are sufficient to keep you from dying after you finish your dinner.

In a world as complex as this one, we have to defer to authorities, and we keep them honest by making those authorities accountable to us and binding them with rules to prevent conflicts of interest. We can't possibly acquire the expertise to adjudicate conflicting claims about the best way to make the world safe and prosperous, but we *can* determine whether the adjudication process itself is trustworthy.

Right now, it's obviously not.

The past 40 years of rising inequality and industry concentration, together with increasingly weak accountability and transparency for expert agencies, has created an increasingly urgent sense of impending doom, the sense that there are vast conspiracies afoot that operate with tacit official approval despite the likelihood they are working to better themselves by ruining the rest of us.

For example, it's been decades since Exxon's own scientists concluded that its products would render Earth uninhabitable by humans. And yet those decades were lost to us, in large part because Exxon lobbied governments and sowed doubt about the dangers of its products and did so with the cooperation of many public officials. When the survival of you and everyone you love is threatened by conspiracies, it's not unreasonable to start questioning the things you think you know in an attempt to determine whether they, too, are the outcome of another conspiracy.

The collapse of the credibility of our systems for divining and upholding truths has left us in a state of epistemological chaos. Once, most of us might have assumed that the system was working and that our regulations reflected our best understanding of the empirical truths of the world as they were best understood—now we have to find our own experts to help us sort the true from the false.

If you're like me, you probably believe that vaccines are safe, but you (like me) probably also can't explain the microbiology or statistics. Few of us have the math skills to review the literature on vaccine safety and describe why their statistical reasoning is sound. Likewise, few of us can review the stats in the (now discredited) literature on opioid safety and explain how those stats were manipulated. Both vaccines and opioids were embraced by medical authorities, after all, and one is safe while the other could ruin your life. You're left with a kind of inchoate constellation of rules of thumb about which experts you trust to fact-check controversial claims and then to explain how all those respectable doctors with their peer-reviewed research on opioid safety *were* an aberration and then how you know that the doctors writing about vaccine safety are *not* an aberration.

I'm 100% certain that vaccinating is safe and effective, but I'm also at something of a loss to explain exactly, *precisely*, why I believe this, given all the corruption I know about and the many times the stamp of certainty has turned out to be a parochial lie told to further enrich the super rich.

Fake news—conspiracy theories, racist ideologies, scientific denialism—has always been with us. What's changed today is not the mix of ideas in the public discourse the popularity of the worst ideas in that mix. Conspiracy and denial have skyrocketed in lockstep with the growth of Big Inequality, which has also tracked the rise of Big Tech and Big Pharma and Big Wrestling and Big Car and Big Movie Theater and Big Everything Else.

No one can say for certain why this has happened, but the two dominant camps are idealism (the belief that the people who argue for these conspiracies have gotten better at explaining them, maybe with the help of machine-learning tools) or materialism (the ideas have become more attractive because of material conditions in the world).

I'm a materialist. I've been exposed to the arguments of conspiracy theorists all my life, and I have not experienced any qualitative leap in the quality of those arguments.

The major difference is in the world, not the arguments. In a time where actual conspiracies are commonplace, conspiracy theories acquire a ring of plausibility.

We have always had disagreements about what's true, but today, we have a disagreement over how we know whether something is true. This is an epistemological crisis, not a crisis over belief. It's a crisis over the credibility of our truth-seeking exercises, from scientific journals (in an era where the biggest journal publishers have been caught producing pay-to-play journals for junk science) to regulations (in an era where regulators are routinely cycling in and out of business) to education (in an era where universities are dependent on corporate donations to keep their lights on).

Targeting—surveillance capitalism—makes it easier to find people who are undergoing this epistemological crisis, but it doesn't create the crisis. For that, you need to look to corruption.

And, conveniently enough, it's corruption that allows surveillance capitalism to grow by dismantling monopoly protections, by permitting reckless collection and retention of personal data, by allowing ads to be targeted in secret, and by foreclosing on the possibility of going somewhere else where you might continue to enjoy your friends without subjecting yourself to commercial surveillance.

Tech is different

I reject both iterations of technological exceptionalism. I reject the idea that tech is uniquely terrible and led by people who are greedier or worse than the leaders of other industries, and I reject the idea that tech is so good—or so intrinsically prone to concentration—that it can't be blamed for its present-day monopolistic status.

I think tech is just another industry, albeit one that grew up in the absence of real monopoly constraints. It may have been first, but it isn't the worst nor will it be the last.

But there's one way in which I *am* a tech exceptionalist. I believe that online tools are the key to overcoming problems that are much more urgent than tech monopolization: climate change; inequality; misogyny; and discrimination on the basis of race, gender identity, and other factors. The internet is how we will recruit people to fight those fights, and how we will coordinate their labor. Tech is not a substitute for democratic accountability, the rule of law, fairness, or stability—but it's a means to achieve these things.

The hard problem of our species is coordination. Everything from climate change to social change to running a business to making a family work can be viewed as a collective action problem.

The internet makes it easier than at any time before to find people who want to work on a project with you—hence the success of free and open-source software, crowdfunding, and racist terror groups—and easier than ever to coordinate the work you do.

The internet and the computers we connect to it also possess an exceptional quality: general-purposeness. The internet is designed to allow any two parties to communicate any data, using any protocol, without permission from anyone else. The only production design we have for computers is the general-purpose, "Turing complete" computer that can run every program we can express in symbolic logic.

This means that every time someone with a special communications need invests in infrastructure and techniques to make the internet faster, cheaper, and more robust, this benefit redounds to everyone else who is using the internet to communicate. And this also means that every time someone with a special computing need invests to make computers faster, cheaper, and more robust, every other computing application is a potential beneficiary of this work.

For these reasons, every type of communication is gradually absorbed into the internet, and every type of device—from airplanes to pacemakers—eventually becomes a computer in a fancy case.

While these considerations don't preclude regulating networks and computers, they do call for gravitas and caution when doing so because changes to regulatory frameworks could ripple out to have unintended consequences in many, many other domains.

The upshot of this is that our best hope of solving the big coordination problems—climate change, inequality, etc.—is with free, fair, and open tech. Our best hope of keeping tech free, fair, and open is to exercise caution in how we regulate tech and to attend closely to the ways in which interventions to solve one problem might create problems in other domains.

Ownership of facts

Big Tech has a funny relationship with information. When you're generating information—anything from the location data streaming off your mobile device to the private messages you send to friends on a social network—it claims the rights to make unlimited use of that data.

But when you have the audacity to turn the tables—to use a tool that blocks ads or slurps your waiting updates out of a social network and puts them in another app that lets you set your own priorities and suggestions or crawls their system to allow you to start a rival business—they claim that you're stealing from them.

The thing is, information is a very bad fit for any kind of private property regime. Property rights are useful for establishing markets that can lead to the effective development of fallow assets. These markets depend on clear titles to ensure that the things being bought and sold in them can, in fact, be bought and sold.

Information rarely has such a clear title. Take phone numbers: There's clearly something going wrong when Facebook slurps up millions of users' address books and uses the phone numbers it finds in them to plot out social graphs and fill in missing information about other users.

But the phone numbers Facebook nonconsensually acquires in this transaction are not the "property" of the users they're taken from nor do they belong to the people whose phones ring when you dial those numbers. The numbers are mere integers, 10 digits in the United States and Canada, and they appear in millions of places, including somewhere deep in pi as well as numerous other contexts. Giving people ownership titles to integers is an obviously terrible idea.

Likewise for the facts that Facebook and other commercial surveillance operators acquire about us, like that we are the children of our parents or the parents to our children or that we had a conversation with someone else or went to a public place. These data points can't be property in the sense that your house or your shirt is your property because the title to them is intrinsically muddy: Does your mom own the fact that she is your mother? Do you? Do both of you? What about your dad—does he own this fact too, or does he have to license the fact from you (or your mom or both of you) in order to use this fact? What about the hundreds or thousands of other people who know these facts?

If you go to a Black Lives Matter demonstration, do the other demonstrators need your permission to post their photos from the event? The online fights over when and how to post photos from demonstrations reveal a nuanced, complex issue that cannot be easily hand-waved away by giving one party a property right that everyone else in the mix has to respect.

The fact that information isn't a good fit with property and markets doesn't mean that it's not valuable. Babies aren't property, but they're inarguably valuable. In fact, we have a whole set of rules just for babies as well as a subset of those rules that apply to humans more generally. Someone who argues that babies won't be truly valuable until they can be bought and sold like loaves of bread would be instantly and rightfully condemned as a monster.

It's tempting to reach for the property hammer when Big Tech treats your information like a nail—not least because Big Tech are such prolific abusers of property hammers when it comes to *their* information. But this is a mistake. If we allow markets to dictate the use of our information, then we'll find that we're sellers in a buyers' market where the Big Tech monopolies set a price for our data that is so low as to be insignificant or, more likely, set at a nonnegotiable price of zero in a click-through agreement that you don't have the opportunity to modify.

Meanwhile, establishing property rights over information will create insurmountable barriers to independent data processing. Imagine that we require a license to be negotiated when a translated document is compared with its original, something Google has done and continues to do billions of times to train its automated language translation tools. Google can afford this, but independent third parties cannot. Google can staff a clearances department to negotiate one-time payments to the likes of the EU (one of the major repositories of translated documents) while independent watchdogs wanting to verify that the translations are well prepared, or to root out bias in translations, will find themselves needing a staffed-up legal department and millions for licenses before they can even get started.

The same goes for things like search indexes of the web or photos of peoples' houses, which have become contentious thanks to Google's Street View project. Whatever problems may exist with Google's photographing of street scenes, resolving them by letting people decide who can take pictures of the facades of their homes from a public street will surely create even worse ones. Think of how street photography is important for newsgathering—including informal newsgathering, like photographing abuses of authority—and how being able to document housing and street life are important for contesting eminent domain, advocating for social aid, reporting planning and zoning violations, documenting discriminatory and unequal living conditions, and more.

The ownership of facts is antithetical to many kinds of human progress. It's hard to imagine a rule that limits Big Tech's exploitation of our collective labors without inadvertently banning people from gathering data on online harassment or compiling indexes of changes in language or simply investigating how the platforms are shaping our discourse—all of which require scraping data that other people have created and subjecting it to scrutiny and analysis.

Persuasion works . . . slowly

The platforms may oversell their ability to persuade people, but obviously, persuasion works sometimes. Whether it's the private realm that LGBTQ people used to recruit allies and normalize sexual diversity or the decadeslong project to convince people that markets are the only efficient way to solve complicated resource allocation problems, it's clear that our societal attitudes *can* change.

The project of shifting societal attitudes is a game of inches and years. For centuries, svengalis have purported to be able to accelerate this process, but even the most brutal forms of propaganda have struggled to make permanent changes. Joseph Goebbels was able to subject Germans to daily, mandatory, hours-long radio broadcasts; to round up and torture and murder dissidents; and to seize full control over their children's education while banning any literature, broadcasts, or films that did not comport with his worldview.

Yet, after 12 years of terror, once the war ended, Nazi ideology was largely discredited in both East and West Germany, and a program of national truth and reconciliation was put in its place. Racism and authoritarianism were never fully abolished in Germany, but neither were the majority of Germans irrevocably convinced of Nazism—and the rise of racist authoritarianism in Germany today tells us that the liberal attitudes that replaced Nazism were no more permanent than Nazism itself.

Racism and authoritarianism have also always been with us. Anyone who's reviewed the kind of messages and arguments that racists put forward today would be hard-pressed to say that they have gotten better at presenting their ideas. The same pseudoscience, appeals to fear, and circular logic that racists presented in the 1980s, when the cause of white supremacy was on the wane, are to be found in the communications of leading white nationalists today.

If racists haven't gotten more convincing in the past decade, then how is it that more people were convinced to be openly racist at that time? I believe that the answer lies in the material world, not the world of ideas. The ideas haven't gotten more convincing, but people have become more afraid. Afraid that the state can't be trusted to act as an honest broker in life-or-death decisions, from those regarding the management of the economy to the regulation of painkillers to the rules for handling private information. Afraid that the world has become a game of musical chairs in which the chairs are being taken away at a never-before-seen rate. Afraid that justice for others will come at their expense. Monopolism isn't the cause of these fears, but the inequality, material desperation and policy malpractice that monopolism contributes to are significant contributors to these conditions. Inequality creates the conditions for both conspiracies and violent racist ideologies, and then surveillance capitalism lets opportunists target the fearful and the conspiracy-minded.

Paying won't help

As the old saw goes, "If you're not paying for the product, you're the product."

It's a commonplace belief today that the advent of free, ad-supported media was the original sin of surveillance capitalism. The reasoning is that the companies that charged for access couldn't "compete with free" and so they were driven out of business. Their ad-supported competitors, meanwhile, declared open season on their users' data in a bid to improve their ad targeting and make more money and then resorted to the most sensationalist tactics to generate clicks on those ads. If only we'd pay for media again, we'd have a better, more responsible, more sober discourse that would be better for democracy.

But the degradation of news products long precedes the advent of ad-supported online news. Long before newspapers were online, lax antitrust enforcement had opened the door for unprecedented waves of consolidation and roll-ups in newsrooms. Rival newspapers were merged, reporters and ad sales staff were laid off, physical plants were sold and leased back, leaving the companies loaded up with debt through leveraged buyouts and subsequent profit-taking by the new owners. In other words, it wasn't merely shifts in the classified advertising market, which was long held to be the primary driver in the decline of the traditional newsroom, that made news companies unable to adapt to the internet—it was monopolism.

Then, as news companies *did* come online, the ad revenues they commanded dropped even as the number of internet users (and thus potential online readers) increased. That shift was a function of consolidation in the ad sales market, with Google and Facebook emerging as duopolists who made more money every year from advertising while paying less and less of it to the publishers whose work the ads appeared alongside. Monopolism created a buyer's market for ad inventory with Facebook and Google acting as gatekeepers.

Paid services continue to exist alongside free ones, and often it is these paid services—anxious to prevent people from bypassing their paywalls or sharing paid media with freeloaders—that exert the most control over their customers. Apple's iTunes and App stores are paid services, but to maximize their profitability, Apple has to lock its platforms so that third parties can't make compatible software without permission. These locks allow the company to exercise both editorial control (enabling it to exclude controversial political material) and technological control, including control over who can repair the devices it makes. If we're worried that ad-supported products deprive people of their right to self-determination by using persuasion techniques to nudge their purchase decisions a few degrees in one direction or the other, then the near-total control a single company holds over the decision of who gets to sell you software, parts, and service for your iPhone should have us very worried indeed.

We shouldn't just be concerned about payment and control: The idea that paying will improve discourse is also dangerously wrong. The poor success rate of targeted advertising means that the platforms have to incentivize you to "engage" with posts at extremely high levels to generate enough pageviews to safeguard their profits. As discussed earlier, to increase engagement, platforms like Facebook use machine learning to guess which messages will be most inflammatory and make a point of shoving those into your eyeballs at every turn so that you will hate-click and argue with people.

Perhaps paying would fix this, the reasoning goes. If platforms could be economically viable even if you stopped clicking on them once your intellectual and social curiosity had been slaked, then they would have no reason to algorithmically enrage you to get more clicks out of you, right?

There may be something to that argument, but it still ignores the wider economic and political context of the platforms and the world that allowed them to grow so dominant.

Platforms are world-spanning and all-encompassing because they are monopolies, and they are monopolies because we have gutted our most important and reliable anti-monopoly rules. Antitrust was neutered as a key part of the project to make the wealthy wealthier, and that project has worked. The vast majority of people on Earth have a negative net worth, and even the dwindling middle class is in a precarious state, undersaved for retirement, underinsured for medical disasters, and undersecured against climate and technology shocks.

In this wildly unequal world, paying doesn't improve the discourse; it simply prices discourse out of the range of the majority of people. Paying for the product is dandy, if you can afford it.

If you think today's filter bubbles are a problem for our discourse, imagine what they'd be like if rich people inhabited free-flowing Athenian marketplaces of ideas where you have to pay for admission while everyone else lives in online spaces that are subsidized by wealthy benefactors who relish the chance to establish conversational spaces where the "house rules" forbid questioning the status quo. That is, imagine if the rich seceded from Facebook, and then, instead of running ads that made money for shareholders, Facebook became a billionaire's vanity project that also happened to ensure that nobody talked about whether it was fair that only billionaires could afford to hang out in the rarified corners of the internet.

Behind the idea of paying for access is a belief that free markets will address Big Tech's dysfunction. After all, to the extent that people have a view of surveillance at all, it is generally an unfavorable one, and the longer and more thoroughly one is surveilled, the less one tends to like it. Same goes for lock-in:

If HP's ink or Apple's App Store were really obviously fantastic, they wouldn't need technical measures to prevent users from choosing a rival's product. The only reason these technical countermeasures exist is that the companies don't believe their customers would *voluntarily* submit to their terms, and they want to deprive them of the choice to take their business elsewhere.

Advocates for markets laud their ability to aggregate the diffused knowledge of buyers and sellers across a whole society through demand signals, price signals, and so on. The argument for surveillance capitalism being a "rogue capitalism" is that machine-learning-driven persuasion techniques distort decision-making by consumers, leading to incorrect signals—consumers don't buy what they prefer, they buy what they're tricked into preferring. It follows that the monopolistic practices of lock-in, which do far more to constrain consumers' free choices, are even more of a "rogue capitalism."

The profitability of any business is constrained by the possibility that its customers will take their business elsewhere. Both surveillance and lock-in are anti-features that no customer wants. But monopolies can capture their regulators, crush their competitors, insert themselves into their customers' lives, and corral people into "choosing" their services regardless of whether they want them—it's fine to be terrible when there is no alternative.

Ultimately, surveillance and lock-in are both simply business strategies that monopolists can choose. Surveillance companies like Google are perfectly capable of deploying lock-in technologies—just look at the onerous Android licensing terms that require device-makers to bundle in Google's suite of applications. And lock-in companies like Apple are perfectly capable of subjecting their users to surveillance if it means keeping the Chinese government happy and preserving ongoing access to Chinese markets. Monopolies may be made up of good, ethical people, but as institutions, they are not your friend—they will do whatever they can get away with to maximize their profits, and the more monopolistic they are, the more they *can* get away with.

An "ecology" moment for trustbusting

If we're going to break Big Tech's death grip on our digital lives, we're going to have to fight monopolies. That may sound pretty mundane and old-fashioned, something out of the New Deal era, while ending the use of automated behavioral modification feels like the plotline of a really cool cyberpunk novel.

Meanwhile, breaking up monopolies is something we seem to have forgotten how to do. There is a bipartisan, trans-Atlantic consensus that breaking up companies is a fool's errand at best—liable to mire your federal prosecutors in decades of litigation—and counterproductive at worst, eroding the "consumer benefits" of large companies with massive efficiencies of scale.

But trustbusters once strode the nation, brandishing law books, terrorizing robber barons, and shattering the illusion of monopolies' all-powerful grip on our society. The trustbusting era could not begin until we found the political will—until the people convinced politicians they'd have their backs when they went up against the richest, most powerful men in the world.

Could we find that political will again?

Copyright scholar James Boyle has described how the term "ecology" marked a turning point in environmental activism. Prior to the adoption of this term, people who wanted to preserve whale populations didn't necessarily see themselves as fighting the same battle as people who wanted to protect the ozone layer or fight freshwater pollution or beat back smog or acid rain.

But the term "ecology" welded these disparate causes together into a single movement, and the members of this movement found solidarity with one another. The people who cared about smog signed petitions circulated by the people who wanted to end whaling, and the anti-whalers marched alongside the people demanding action on acid rain. This uniting behind a common cause completely changed the dynamics of environmentalism, setting the stage for today's climate activism and the sense that preserving the habitability of the planet Earth is a shared duty among all people.

I believe we are on the verge of a new "ecology" moment dedicated to combating monopolies. After all, tech isn't the only concentrated industry nor is it even the *most* concentrated of industries.

You can find partisans for trustbusting in every sector of the economy. Everywhere you look, you can find people who've been wronged by monopolists who've trashed their finances, their health, their privacy, their educations, and the lives of people they love. Those people have the same cause as the people who want to break up Big Tech and the same enemies. When most of the world's wealth is in the hands of a very few, it follows that nearly every large company will have overlapping shareholders.

That's the good news: With a little bit of work and a little bit of coalition building, we have more than enough political will to break up Big Tech and every other concentrated industry besides. First we take Facebook, then we take AT&T/WarnerMedia.

But here's the bad news: Much of what we're doing to tame Big Tech *instead* of breaking up the big companies also forecloses on the possibility of breaking them up later.

Big Tech's concentration currently means that their inaction on harassment, for example, leaves users with an impossible choice: absent themselves from public discourse by, say, quitting Twitter or endure vile, constant abuse. Big Tech's over-collection and over-retention of data results in horrific identity theft. And their inaction on extremist recruitment means that white supremacists who livestream their shooting rampages can reach an audience of billions. The combination of tech concentration and media concentration means that artists' incomes are falling even as the revenue generated by their creations are increasing.

Yet governments confronting all of these problems all inevitably converge on the same solution: deputize the Big

Tech giants to police their users and render them liable for their users' bad actions. The drive to force Big Tech to use automated filters to block everything from copyright infringement to sex-trafficking to violent extremism means that tech companies will have to allocate hundreds of millions to run these compliance systems.

These rules—the EU's new Directive on Copyright, Australia's new terror regulation, America's FOSTA/SESTA sex-trafficking law and more—are not just death warrants for small, upstart competitors that might challenge Big Tech's dominance but who lack the deep pockets of established incumbents to pay for all these automated systems. Worse still, these rules put a floor under how small we can hope to make Big Tech.

That's because any move to break up Big Tech and cut it down to size will have to cope with the hard limit of not making these companies so small that they can no longer afford to perform these duties—and it's *expensive* to invest in those automated filters and outsource content moderation. It's already going to be hard to unwind these deeply concentrated, chimeric behemoths that have been welded together in the pursuit of monopoly profits. Doing so while simultaneously finding some way to fill the regulatory void that will be left behind if these self-policing rulers were forced to suddenly abdicate will be much, much harder.

Allowing the platforms to grow to their present size has given them a dominance that is nearly insurmountable— deputizing them with public duties to redress the pathologies created by their size makes it virtually impossible to reduce that size. Lather, rinse, repeat: If the platforms don't get smaller, they will get larger, and as they get larger, they will create more problems, which will give rise to more public duties for the companies, which will make them bigger still.

We can work to fix the internet by breaking up Big Tech and depriving them of monopoly profits, or we can work to fix Big Tech by making them spend their monopoly profits on governance. But we can't do both. We have to choose between a vibrant, open internet or a dominated, monopolized internet commanded by Big Tech giants that we struggle with constantly to get them to behave themselves.

Make Big Tech small again

Trustbusting is hard. Breaking big companies into smaller ones is expensive and time-consuming. So time-consuming that by the time you're done, the world has often moved on and rendered years of litigation irrelevant. From 1969 to 1982, the U.S. government pursued an antitrust case against IBM over its dominance of mainframe computing—but the case collapsed in 1982 because mainframes were being speedily replaced by PCs.

It's far easier to prevent concentration than to fix it, and reinstating the traditional contours of U.S. antitrust enforcement will, at the very least, prevent further concentration. That means bans on mergers between large companies, on big companies acquiring nascent competitors, and on platform companies competing directly with the companies that rely on the platforms.

These powers are all in the plain language of U.S. antitrust laws, so in theory, a future U.S. president could simply direct their attorney general to enforce the law as it was written. But after decades of judicial "education" in the benefits of monopolies, after multiple administrations that have packed the federal courts with lifetime-appointed monopoly cheerleaders, it's not clear that mere administrative action would do the trick.

If the courts frustrate the Justice Department and the president, the next stop would be Congress, which could eliminate any doubt about how antitrust law should be enforced in the United States by passing new laws that boil down to saying, "Knock it off. We all know what the Sherman Act says. Robert Bork was a deranged fantasist. For avoidance of doubt, *fuck that guy*." In other words, the problem with monopolies is *monopolism*—the concentration of power into too few hands, which erodes our right to self-determination. If there is a monopoly, the law wants it gone, period. Sure, get rid of monopolies that create "consumer harm" in the form of higher prices, but also, *get rid of other monopolies, too*.

But this only prevents things from getting worse. To help them get better, we will have to build coalitions with other activists in the anti-monopoly ecology movement—a pluralism movement or a self-determination movement—and target existing monopolies in every industry for breakup and structural separation rules that prevent, for example, the giant eyewear monopolist Luxottica from dominating both the sale and the manufacture of spectacles.

In an important sense, it doesn't matter which industry the breakups begin in. Once they start, shareholders in *every* industry will start to eye their investments in monopolists skeptically. As trustbusters ride into town and start making lives miserable for monopolists, the debate around every corporate boardroom's table will shift. People within corporations who've always felt uneasy about monopolism will gain a powerful new argument to fend off their evil rivals in the corporate hierarchy: "If we do it my way, we make less money; if we do it your way, a judge will fine us billions and expose us to ridicule and public disapprobation. So even though I get that it would be really cool to do that merger, lock out that competitor, or buy that little company and kill it before it can threaten it, we really shouldn't—not if we don't want to get tied to the DOJ's bumper and get dragged up and down Trustbuster Road for the next 10 years."

20 GOTO 10

Fixing Big Tech will require a lot of iteration. As cyber lawyer Lawrence Lessig wrote in his 1999 book, *Code and Other Laws of Cyberspace*, our lives are regulated by four forces: law (what's legal), code (what's technologically possible), norms (what's socially acceptable), and markets (what's profitable).

If you could wave a wand and get Congress to pass a law that re-fanged the Sherman Act tomorrow, you could use the impending breakups to convince venture capitalists to fund competitors to Facebook, Google, Twitter, and Apple that would be waiting in the wings after they were cut down to size.

But getting Congress to act will require a massive normative shift, a mass movement of people who care about monopolies—and pulling them apart.

Getting people to care about monopolies will take technological interventions that help them to see what a world free from Big Tech might look like. Imagine if someone could make a beloved (but unauthorized) third-party Facebook or Twitter client that dampens the anxiety-producing algorithmic drumbeat and still lets you talk to your friends without being spied upon—something that made social media more sociable and less toxic. Now imagine that it gets shut down in a brutal legal battle. It's always easier to convince people that something must be done to save a thing they love than it is to excite them about something that doesn't even exist yet.

Neither tech nor law nor code nor markets are sufficient to reform Big Tech. But a profitable competitor to Big Tech could bankroll a legislative push; legal reform can embolden a toolsmith to make a better tool; the tool can create customers for a potential business who value the benefits of the internet but want them delivered without Big Tech; and that business can get funded and divert some of its profits to legal reform. 20 GOTO 10 (or lather, rinse, repeat). Do it again, but this time, get farther! After all, this time you're starting with weaker Big Tech adversaries, a constituency that understands things can be better, Big Tech rivals who'll help ensure their own future by bankrolling reform, and code that other programmers can build on to weaken Big Tech even further.

The surveillance capitalism hypothesis—that Big Tech's products really work as well as they say they do and that's why everything is so screwed up—is way too easy on surveillance and even easier on capitalism. Companies spy because they believe their own BS, and companies spy because governments let them, and companies spy because any advantage from spying is so short-lived and minor that they have to do more and more of it just to stay in place.

As to why things are so screwed up? Capitalism. Specifically, the monopolism that creates inequality and the inequality that creates monopolism. It's a form of capitalism that rewards sociopaths who destroy the real economy to inflate the bottom line, and they get away with it for the same reason companies get away with spying: because our governments are in thrall to both the ideology that says monopolies are actually just fine and in thrall to the ideology that says that in a monopolistic world, you'd better not piss off the monopolists.

Surveillance doesn't make capitalism rogue. Capitalism's unchecked rule begets surveillance. Surveillance isn't bad because it lets people manipulate us. It's bad because it crushes our ability to be our authentic selves—and because it lets the rich and powerful figure out who might be thinking of building guillotines and what dirt they can use to discredit those embryonic guillotine-builders before they can even get to the lumberyard.

Up and through

With all the problems of Big Tech, it's tempting to imagine solving the problem by returning to a world without tech at all. Resist that temptation.

The only way out of our Big Tech problem is up and through. If our future is not reliant upon high tech, it will be because civilization has fallen. Big Tech wired together a planetary, species-wide nervous system that, with the proper reforms and course corrections, is capable of seeing us through the existential challenge of our species and planet. Now it's up to us to seize the means of computation, putting that electronic nervous system under democratic, accountable control.

I am, secretly, despite what I have said earlier, a tech exceptionalist. Not in the sense of thinking that tech should be given a free pass to monopolize because it has "economies of scale" or some other nebulous feature. I'm a tech exceptionalist because I believe that getting tech right matters and that getting it wrong will be an unmitigated catastrophe—and doing it right can give us the power to work together to save our civilization, our species, and our planet.

Surveillance Capitalism 'n Kids

Give me the child for the first seven years, and I will give you the man.

The masses have been panicking over those damn kids and their new media for a *long* time. Novels were thought to be poisoning impressionable young women's minds, the waltz a gateway to a life of sin and depredation, *Dungeons and Dragons* a stalking horse for satanism.

Is the internet any different?

Yes . . . and no. Some of the handwringing over kids' use of screens and services has the stink of a moral panic. When hoaxes like the Momo Challenge are uncritically taken up and repeated by teachers, parenting advice pundits, public safety officials, and others, you know that at least some of the concern over kids' tech use is driven by the same old impulses that allegedly prompted Socrates to harrumph that "The children now love luxury; they have bad manners, contempt for authority; they show disrespect for elders and love chatter in place of exercise. Children are now tyrants, not the servants of their households."

The narrative about "digital natives" doesn't help. Kids are not born with an innate sense of how technology should work; at best, their lack of preconceptions inherited from mastery of older forms of media means that they grasp some aspects of a new technology a little more quickly. And of course, kids can be first-rate reasoners with the ability to build up understandings from first principles.

But what kids lack is context. Kids sometimes become math or chess prodigies because you can learn the ground truths for those fields in a short while and then apply whatever reason you have to understanding and mastering them. But kids don't become history or law prodigies because there just aren't enough hours in the day for a kid to do the reading one needs to do in order to start applying reason to knowledge of the subject.

That means that kids are naive. They have to learn—often the hard way—all the ways that seemingly trustworthy media can be manipulative. Just as your immune system needs to contend with and overcome a long procession of pathogens before you stop getting "childhood illnesses," so too does your cognitive immune system need to contend with a wide variety of underhanded and manipulative techniques before you will be able to automatically resist them.

Therein lies the problem. Our cognitive immune systems and the persuasion researchers of surveillance capitalism are locked in their Red Queen's race, each driving the evolution of the other. In a very short time, our species' arsenal of manipulative techniques and countermeasures has grown in sophistication and variety in a way that staggers the imagination (recall that the first banner ad had a click-through rate of 44% in 1994; today, if you can find a banner ad, you can be sure it's attracting a click-through rate well below 0.2%).

This highlights another way in which "digital native" rhetoric is unhelpful. Not only are kids not born with an innate understanding of how technology is supposed to work, but they are also born without having lived through any of the surveillance capitalism evolutionary arms race. They enter an internet that has had decades to evolve powerful—if short-lived—manipulation techniques while they have had no time to evolve the countermeasures the rest of us deploy automatically. They are newborns with newborn cognitive immune systems, and we have selectively bred a menagerie of persuasion superbugs; as soon as they can hold a screen, we unleash these bugs on them.

Kids do need parental supervision when using networks. Filters can help, but they overblock (catching things that kids should be able to see, especially age-appropriate material related to human sexuality and sexual health) and underblock (recall that YouTube Kids has been repeatedly hijacked by disturbing, violent videos as well as sleazy and deceptive "influencer marketing").

It's tempting to think of kids as having technology superpowers, and there is one sense in which they do:

Kids are powerfully motivated to investigate novel ways of communicating with their peers—both their friends and strangers around the world. In part, that's because of the intensely social developmental process, but it's also a function of the disappearance of public spaces in which kids are welcome and spaces in which kids can autonomously socialize with peers without structure and without parental oversight.

The end of kids' freedom to move through our cities and towns—a combination of statistically unfounded fear of "stranger danger" and decades of underinvestment in public transit—has left kids with overstructured lives in which all of their socializing is in the context of highly regimented school and extracurricular activities.

Faced with these barriers to the normal socializing that is key to development, kids are willing to put enormous energy into finding ways to be sociable with one another, seeking out games, social media, and other online environments that can fill in for those lost, unstructured "hanging out" opportunities.

But it would be a mistake to confuse willingness to adopt a technology with an aptitude for early adoption. Given enough time and energy, any of us can master a new medium or communications tool, but most of us don't need to because the existing tools work just fine for us. Why bother to learn to use Snapchat if Facebook is filling your needs? But if you're a kid looking to escape your parents' monitoring of Facebook, learning Snapchat requires less effort than figuring out how to avoid your parents on Facebook.

We often mix up a need for new technologies with an affinity for new technologies. For example, there's a commonly held view that the pornography industry is intrinsically technophilic, the first to adopt new technologies, and there's certainly lots of evidence to support this thesis: Porn was early in photography, 8mm film, home videos, and the internet (to say nothing of cryptocurrency and other alternative payment schemes).

But there's no obvious link between the carnal and the technological. Rather, there's another connection: When you traffic in banned or disfavored material like pornography, merely communicating requires an enormous effort that people in the mainstream do not have to contend with. If it's 1950 and you've made a commercially viable film, some cinemas will exhibit it, but if it's an "adult film," you will have to arrange a secret showing in a makeshift theater and hope your audience will discover it (and that the vice squad won't).

For people whose communications are not censored, learning to use a VCR or the internet is an unnecessary cost. For those whose communications are officially disfavored—terrorists, heretics, pornographers, children—all the costs associated with mastering a new medium are a bargain.

So kids aren't technophiles any more than pornographers or terrorists are. They just get benefits out of using technology that the rest of us take for granted: the power to have unfettered, unmonitored communications.

Kids' lack of opportunity to learn to identify and avoid the cognitive traps laid by surveillance capitalists combined with their burning need to connect with one another and the pressures that drive them to seek out new online spaces beyond parental awareness and control poses a unique danger.

But this danger is not because of "what screens are doing to our kids' brains." We have a unique and thoroughly unscientific horror of our brains being changed, but another word for "changing your brain" is "learning." Cognition happens in our brains, and cognition is a matter of neuronal firing and wiring, such that every thought can be said to be "changing your brain."

Learning different things affects our brains in different ways, and it may be that using, say, an iPad interface between the ages of three and five will produce a different effect to sitting in front of a cable TV remote during those years, but 1) we won't be able to meaningfully determine what those differences are until those iPad kids are adults, and 2) by then, iPads will no longer exist. (My thanks to UC Irvine's Mizuka Ito for this insight.)

The problem with kids in a world of surveillance capitalism isn't that their brains are being rewired. It's that they might be tricked into harmful activities and beliefs, including some that might endure into adulthood.

Ingram Content Group UK Ltd.
Milton Keynes UK
UKHW021304060623
422961UK00024B/856